Coimbra Mathematical Texts

Volume 2

The Mathematics Department of the University of Coimbra, one of the oldest universities in the world, has a new collection of advanced level mathematical texts, *Coimbra Mathematical Texts*, in collaboration with Springer Nature. This collection aims to publish monographs and conference proceedings of excellent quality, in all branches of mathematics, that will inspire the future generations.

Any researcher interested to contribute to this collection should send an e-mail to cmt@mat.uc.pt with subject "Possible contribution".

The editorial team will perform an initial appraisal of the submitted manuscript based on the quality of the submission, interest and importance of the topic. If the contribution is considered suitable to be sent to peer review, it will be reviewed by specialists of international reputation.

Francis Borceux

Galois Theories of Fields and Rings

 Springer

Francis Borceux
Department of Mathematics
University of Louvain
Louvain-la-Neuve, Belgium

ISSN 2813-0057 ISSN 2813-0065 (electronic)
Coimbra Mathematical Texts
ISBN 978-3-031-58459-6 ISBN 978-3-031-58460-2 (eBook)
https://doi.org/10.1007/978-3-031-58460-2

This Springer imprint is published by the registered company Springer Nature Switzerland AG
The registered company address is: Gewerbestrasse 11, 6330 Cham, Switzerland

If disposing of this product, please recycle the paper.

To Rose
with love

Preface to the Book Series
Coimbra Mathematical Texts

This new Springer series "Coimbra Mathematical Texts" is a direct successor of a collection entitled "Textos de Matemática", published by the University of Coimbra, which started in 1993.

The aim of that collection was the publication of advanced mathematics texts resulting from events held at the University of Coimbra—Mathematics Department, and included conference proceedings as well as monographs related to graduate-level short courses. The first volume in the collection was "Classical Invariants", by J. A. Green (1993), and the last one, no. 48, was "13th Young Researchers Workshop on Geometry, Mechanics and Control: three mini-courses" (2019). The complete list of volumes, covering a wide range of mathematical topics, may be found at

https://www.uc.pt/fctuc/dmat/seccoes/publicacoes/textosDeMatematica

All volumes went through a rigorous refereeing process.

The series now gains a new life with the launching of "Coimbra Mathematical Texts", a partnership between Springer Nature and the University of Coimbra. The new collection maintains the same spirit but aims at publishing advanced texts from more diverse origins. We therefore invite the wider mathematical community to submit quality contributions to the series, both monographs and proceedings, in all branches of mathematics.

Inquiries regarding the submission process should be sent to cmt@mat.uc.pt.

Ana Paula Santana
Júlio S. Neves
Marcelo Viana
Maria Paula Serra de Oliveira
Rui Loja Fernandes
The CMT Series Editors

Preface

This book intends to make the Galois theory of rings accessible to a wide audience interested in ring theory or category theory. The presentation starts with the classical Galois theory of fields and eventually reaches a Galois theory of commutative rings. The key to this generalization is the work of A. Grothendieck, which extends the classical Galois bijection for fields to an equivalence of categories involving split algebras and profinite topological spaces.

The category theory community's interest in generalized Galois theories, including the topics of the present book, goes back to the International Category Theory Meeting held in Louvain-la-Neuve (Belgium) in 1987. At that meeting, Saunders Mac Lane delivered a talk on recent work of the Georgian mathematician George Janelidze, who could not attend the meeting due to the political situation at that time. The talk was about a very general notion of Galois descent in the context of some well-behaved adjunctions.

In the subsequent years, a seminar was held in Louvain-la-Neuve in order to exhibit clearly, step by step, the various generalizations which allow one to pass progressively from the classical Galois theory of fields to the very general abstract setting of Galois descent studied by George Janelidze. The never-published set of notes of this seminar is at the origin of the book *Galois Theories*, by F. Borceux and G. Janelidze, published by Cambridge University Press in 2001 (see [7]).

The content of the present book is based on a master's course on Galois theory given by the author at the University of Coimbra (Portugal) in 2003. The audience had a good background in category theory. The notes of this course were published in 2004 as the volume *Algumas Teorias de Galois dos Corpos e dos Aneís*, in the *Textos de Matemática* series of the University of Coimbra (see [8]). Regarding the corresponding treatment of the same topics in the Borceux–Janelidze book, the point of view followed in the Portuguese volume takes advantage of new results of B. Mesablishvili (see [32]), proving that the effective descent morphisms of rings are exactly the pure ones.

This English translation of the Portuguese volume offered the opportunity to introduce various improvements. But more importantly, special attention has been devoted to making the book as self-contained as possible by recalling all non-elementary notions of ring theory and category theory which are used. This has been done by extending existing sections, introducing new sections, or just by adding footnotes in the case of more punctual aspects.

The author thanks Maria Manuel Clementino, who invited him to teach the master's course at the University of Coimbra which is at the origin of this book. Many thanks also to Lurdes Sousa, whose careful reading improved the quality of the text, and to Júlio Neves for his efficient technical help.

Université Catholique de Louvain   *Francis Borceux*
November 2023   francis.borceux@uclouvain.be

Contents

Part II The Galois Theory of Rings

Historical Introduction

This unusually long introduction is intended to be both a short history of the considerations which, progressively, led to the development of Galois theory, but also a detailed overview of the content of this book. So the reader who just wants to discover and understand – without any proofs – what the various Galois theories of fields and rings are about, and where they come from, can simply refer to this introduction.

The whole book (except this introduction) is intended to be as self-contained as possible. Only some elementary familiarity is assumed with various basic notions from the theories of groups (including quotients), rings (including ideals), fields, vector spaces, modules (including tensor products), Boolean algebras, categories (including naturality, duality) and topological spaces (including Hausdorffness, compactness). The more involved aspects of these various theories are explicitly recalled.

The first part of this book presents various Galois theorems for fields:

- the classical Galois theorem for finite-dimensional field extensions;
- its generalization in terms of finite-dimensional split algebras;
- the Galois theorem for arbitrary field extensions;
- its generalization in terms of arbitrary split algebras.

It is this last theorem which shows the way to a Galois theorem for rings: the topic of the second part of this book.

The equations of the first and second degree

Everything started with the problem of solving equations. Already 15 centuries before Christ, mathematicians were able to solve the linear equation

$$aX + b = 0, \quad X = -\frac{b}{a},$$

as well as the equation of the second degree

$$aX^2 + bX + c = 0, \quad X = \frac{-b \pm \sqrt{b^2 - 4ac}}{2a}.$$

© The Author(s), under exclusive license to Springer Nature Switzerland AG 2024
F. Borceux, *Galois Theories of Fields and Rings*, Coimbra Mathematical Texts 2,
https://doi.org/10.1007/978-3-031-58460-2_1

Of course in those days, things were not written in this formal algebraic form and the solution methods were often more geometrical than algebraic. Also, for many centuries, every number considered was a real (positive) number ... even if that notion had not yet been made precise as it is today. In particular, the quantity $\sqrt{b^2 - 4ac}$ could not create any problems, because its existence is precisely equivalent to the existence of a real solution.

The equations of the third and fourth degree

The Italian mathematicians of the 16th century (Tartaglia, del Fiore, Ferrari, Cardano, ...) found formulæ for solving the equations of the third and fourth degree. For example, putting

$$Y = X + \frac{b}{3a}$$

in the equation

$$aX^3 + bX^2 + cX + d = 0,$$

yields an equation of the form

$$Y^3 + pY + q = 0.$$

This equation admits the root

$$Y = \sqrt[3]{-\frac{q}{2} + \sqrt{\left(\frac{p}{3}\right)^3 + \left(\frac{q}{2}\right)^2}} + \sqrt[3]{-\frac{q}{2} - \sqrt{\left(\frac{p}{3}\right)^3 + \left(\frac{q}{2}\right)^2}}.$$

In the special case of the equation

$$Y^3 - 15Y - 4 = 0,$$

which admits the three real roots

$$Y_1 = 4, \quad Y_2, Y_3 = -2 \pm \sqrt{3},$$

the formula above produces

$$Y = \sqrt[3]{2 + \sqrt{-121}} + \sqrt[3]{2 - \sqrt{-121}},$$

which contains square roots of negative numbers. A highly puzzling new challenge for the mathematicians of those days! But they realized that by accepting "imaginary numbers" like the square roots of negative numbers, and assuming that they behave as "actual numbers", they were able to compute that the formula above yields the root

$$Y = 4.$$

The solution of the equation of third degree thus led to the consideration of a new set of numbers, which we now call the *complex numbers*. Today, nobody still uses the formulæ for solving the equations of the third and the fourth degree[1] ... but complex numbers have become an essential and unavoidable notion in mathematics.

The equations of degree five and more

Three more centuries would pass before an answer was found in the case of the equation of degree five, or more. The mathematicians Abel and Galois, independently between 1824 and 1830, found necessary and sufficient conditions for the existence of a formula, with radicals, for solving a polynomial equation of arbitrary degree. They pointed out in particular some equations of the fifth degree which do not satisfy their conditions, and thus which cannot be solved via a formula applied to the coefficients of the equation and using additions, subtractions, multiplications, divisions and radicals. So, there cannot exist a general formula for solving equations of degree five or more.

Like in the case of the equation of the third degree and the discovery of complex numbers, we no longer use the conditions discovered by Abel and Galois to determine if we can solve an equation of higher degree by radicals. But the notions that they introduced to achieve their goal remain fundamental today: the notions of group, field, and of course, the so-called *Galois theory*.

To study the possibility of solving by radicals an equation of the form

$$p(X) = a_n X^n + a_{n-1} X^{n-1} + \cdots + a_1 X + a_0, \quad a_i \in \mathbb{C}$$

Galois considers on one hand the subfield $K \subseteq \mathbb{C}$ generated by the coefficients a_i of the equation, and on the other hand the subfield $L \subseteq \mathbb{C}$ generated by the coefficients and the roots. By definition of the field L, the polynomial $p(X)$ admits a decomposition in linear factors[2] over the field L. The extension $K \subseteq L$ has the striking property that each element $l \in L$ is root of a polynomial $q(X) \in K[X]$ which factors into distinct polynomials of degree 1 over L. Such an extension is called a *Galois extension*.

In the case of the equation

$$Y^3 - 15Y - 4 = 0$$

already mentioned, $K = \mathbb{Q}$ and $L \subseteq \mathbb{C}$ is the subfield of $\mathbb{C}$ generated by Y_2 (or equivalently, Y_3). More precisely,

$$L = \left\{ u + v\sqrt{3} \,\middle|\, u, v \in \mathbb{Q} \right\}.$$

---

[1] Instead we use iterative methods, like the *Newton–Raphson* method, which allow us to compute the root of a polynomial (or even more general) equation $p(X) = 0$ of arbitrary degree, with "as many decimals as wanted".

[2] Galois considers the case of distinct factors; this is not a severe restriction because every polynomial $p(X) \in K[X]$ is a product of polynomials $p_i(X) \in K[X]$ such that each $p_i(X) \in K[X]$ has distinct roots in $\mathbb{C}$.

Let us observe that indeed, L is field, because

$$\frac{1}{u + v\sqrt{3}} = \frac{-u}{3v^2 - u^2} + \frac{v}{3v^2 - u^2}\sqrt{3}$$

and $3v^2 - u^2 = 0$ is impossible if u, v are rational numbers.

Each field homomorphism $f : L \longrightarrow L$ is such that $f(1) = 1$ and, for this reason, fixes all elements of $\mathbb{Q} = K$:

$$\forall k \in K = \mathbb{Q} \quad f(k) = k.$$

Moreover, f is entirely determined by the value of $f(\sqrt{3})$. But

$$\left(f(\sqrt{3})\right)^2 = f\left((\sqrt{3})^2\right) = f(3) = 3$$

and so we conclude that

$$f(\sqrt{3}) = \sqrt{3} \quad \text{or} \quad f(\sqrt{3}) = -\sqrt{3}.$$

In the first case, f is the identity on L; in the second case, f is a field homomorphism such that

$$f(Y_2) = Y_3, \quad f(Y_3) = Y_2.$$

Thus the field homomorphisms $f : L \longrightarrow L$ fixing all the elements of K are determined by the permutations of the roots. These homomorphisms are isomorphisms and, for this reason, constitute a group: this group is called the *Galois group* of the equation.

Galois proved that the possibility of solving the equation $p(X) = 0$ by radicals depends on properties of the group of permutation of the roots, more precisely, of the properties of the subgroups of the Galois group.

The classical Galois theory

More generally, an extension $K \subseteq L$ of fields is called a *Galois extension* when each element $l \in L$ is root of a polynomial $q(X) \in K[X]$ which admits a decomposition into distinct linear factors over L. The Galois group $\mathsf{Gal}[L : K]$ of the extension is the group of field automorphisms $f : L \longrightarrow L$ fixing all the elements of K.

The classical Galois theorem asserts that when $K \subseteq L$ is a finite-dimensional Galois extension, there exists a bijective correspondence between the subgroups $G \subseteq \mathsf{Gal}[L : K]$ of the Galois group and the intermediate field extensions $K \subseteq M \subseteq L$. Moreover, the cardinality of the Galois group $\mathsf{Gal}[L : K]$ is equal to the dimension of L as a K-vector space.

The Grothendieck Galois theorem

There exists a generalization of this classical result, due to Grothendieck. If $K \subseteq L$ is a field extension, a K-algebra A is said to be *split over L* when each element $a \in A$ is a root of a polynomial $q(X) \in K[X]$ which admits a decomposition into distinct linear factors over L. The extension $K \subseteq L$ is thus a *Galois extension* when the field L, viewed as a K-algebra, is split over L.

When $K \subseteq L$ is a Galois extension of finite dimension n, we know that the Galois group $\mathsf{Gal}[L : K]$ is finite, with n elements. But for each finite-dimensional K-algebra A split over L, the set $\mathsf{Hom}_K(A, L)$ of K-linear homomorphisms is finite as well. Moreover, composition of homomorphisms yields an action of the Galois group on $\mathsf{Hom}(A, L)$:

$$\mathsf{Gal}[L : K] \times \mathsf{Hom}_K(A, L) \longrightarrow \mathsf{Hom}_K(A, L), \quad (g, f) \mapsto g \circ f.$$

The Grothendieck Galois theorem asserts that when $K \subseteq L$ is a finite-dimensional Galois extension, the mapping

$$A \mapsto \mathsf{Hom}_K(A, L)$$

defines a contravariant equivalence of categories between the category of finite-dimensional K-algebras split by L, and the category of finite sets provided with an action of the Galois group $\mathsf{Gal}[L : K]$. In particular, this equivalence reduces as a bijection between the intermediate subextensions $K \subseteq M \subseteq L$ and the quotients of $\mathsf{Hom}_K(L, L) = \mathsf{Gal}[L : K]$. This last observation recaptures the classical Galois theorem, because given a group G, the quotients of the G-set G are in one-to-one correspondence with the subgroups of G.

The Galois theorems in infinite dimension

It is possible to generalize both Galois theorems already mentioned to the case of a Galois extension $K \subseteq L$ of arbitrary dimension. To achieve this, it is necessary to involve topological arguments.

A Galois extension $K \subseteq L$ of arbitrary dimension is always the filtered union – thus a filtered colimit – of the finite-dimensional intermediate Galois extensions $K \subseteq M \subseteq L$. This yields an isomorphism

$$\mathsf{Gal}[L : K] \cong \lim \mathsf{Gal}[M : K], \quad g \mapsto (g|_M)_M$$

where the limit is now a cofiltered one. One puts the discrete topology on each finite Galois group $\mathsf{Gal}[M : K]$; these are thus in particular (trivial) compact Hausdorff spaces. Next, one puts the corresponding limit topology in Top on $\mathsf{Gal}[L : K]$. A topological space obtained in this way is called *profinite*. It is a compact Hausdorff space possessing a base of closed open subsets.

In the same way, a K-algebra A split by L is the union of all the finite-dimensional sub-K-algebras $B \subseteq A$, split by L. This yields an inclusion

$$\mathsf{Hom}_K(A, L) \rightarrowtail \prod_B \mathsf{Hom}_K(B, L)$$

and choosing the discrete topology on each $\mathsf{Hom}(B, L)$, the product topology and the induced topology, we also provide $\mathsf{Hom}_K(A, L)$ with a profinite topology.

This makes it possible to generalize both Galois theorems mentioned above:

- there exists a bijection between the closed subgroups $G \subseteq \mathrm{Gal}[L : K]$ of the profinite Galois group and the intermediate field extensions $K \subseteq M \subseteq L$;
- there exists a contravariant equivalence of categories

$$A \mapsto \mathrm{Hom}_K(A, L)$$

between the category of K-algebras split by L and the category of profinite topological spaces provided with a continuous action of the profinite Galois group.

This last result is not only an important generalization of the classical Galois theorem, it is also the key which opens the door to a Galois theorem for rings, where no notion of dimension exists.

Pure extensions of rings

In this book all rings considered are commutative and with a unit. Their Galois theory requires some less known notions of ring theory.

First, in the case of rings, a Galois extension $R \subseteq S$ must be *pure* ... a condition which does not appear in the case of fields, because every field extension is pure. A morphism $f \colon A \longrightarrow B$ of R-modules is pure when, for each R-module M, the morphism

$$M \otimes_R f \colon M \otimes_R A \longrightarrow M \otimes_R B$$

is injective. In particular, the morphism f itself is injective (case $M = R$). The extension $R \subseteq S$ is then *pure* when the inclusion is a pure morphism in the category of R-modules.

When K is a field, each K-linear injection $f \colon A \rightarrowtail B$ admits a restriction $s \colon B \longrightarrow\!\!\!\!\!\rightarrow A$; as a consequence, $M \otimes_K f$ admits the restriction $M \otimes_K s$ and therefore, is injective. Thus every K-linear injection f is pure.

Descent theory of rings

Pure extensions of rings are important because they are exactly the descent morphisms of the theory of rings: an essential key to proving a Galois theorem.

Each ring homomorphism $f \colon R \longrightarrow S$ induces a pair of adjoint functors between the corresponding categories of modules

$$\mathrm{Mod}_S \underset{U}{\overset{S \otimes_R -}{\rightleftarrows}} \mathrm{Mod}_R, \quad S \otimes_R - \dashv U.$$

The functor $U(M) = M$ is always monadic. This means that the composite functor

$$T \colon \mathrm{Mod}_R \longrightarrow \mathrm{Mod}_R, \quad M \mapsto S \otimes_R M$$

is provided with a unit and a multiplication

$$\varepsilon \colon \mathrm{id}_{\mathrm{Mod}_R} \Rightarrow T, \quad \mu \colon T \circ T \Rightarrow T.$$

which are defined by

$$\varepsilon_M : M \longrightarrow S \otimes_R M, \quad m \mapsto 1 \otimes m,$$

$$\mu_M : S \otimes_R S \otimes_R M \longrightarrow S \otimes_R M, \quad s \otimes s' \otimes m \mapsto ss' \otimes m,$$

and satisfy axioms analogous to those for a monoid. Saying that U is monadic means that the category of S-modules is equivalent to the category of those pairs (M, ξ) of an R-module M together with an adequate action ξ

$$M \in \mathsf{Mod}_R \quad \text{with} \quad \xi : T(M) \longrightarrow M \quad \text{in } \mathsf{Mod}_R.$$

Indeed, an S-module is an R-module provided with an R-linear action

$$S \otimes_R M \longrightarrow M, \quad s \otimes m \mapsto sm.$$

The ring homomorphism $f : R \longrightarrow S$ is a *descent morphism* when moreover, the functor $S \otimes_R -$ is co-monadic, that is, has the dual property. In other words, when each R-module can be seen as an S-module provided with an adequate co-action. It turns out that the descent morphisms of rings are exactly the pure (mono)morphisms of rings.

The spectrum of a ring

A second notion that we must introduce is the *spectrum of a ring*. This is a topological space associated with the ring. There are various non-equivalent ways of associating a topological space with a ring: the one which is useful for Galois theory is the so-called *Pierce spectrum*, which is a profinite space. Again this notion does not appear in the case of fields, because the Pierce spectrum of a field is just a singleton.

An element $e \in S$ of a ring is *idempotent* when $e^2 = e$. The operations

$$e \wedge e' = ee', \quad e \vee e' = e + e' - ee'$$

provide the set of idempotents of the ring S with the structure of a Boolean algebra.

The Stone duality theorem asserts that the category of Boolean algebras is the dual of the category of profinite topological spaces. The Pierce spectrum of a ring is the profinite space corresponding, by the Stone duality, to its Boolean algebra of idempotents.

The Pierce spectrum also admits a direct description. An ideal $I \triangleleft S$ is *regular* when it is generated by its idempotent elements. The Pierce spectrum $\mathsf{Spec}(S)$ of S is the set of its maximal regular ideals, provided with the topology whose open subsets are the

$$O_I = \big\{ M \in \mathsf{Spec}(S) \big| I \not\subseteq M \big\},$$

for all regular ideals $I \triangleleft S$.

In the case of a field L, the only idempotents are 0 and 1 and the only (regular) ideals are $\{0\}$ and L. Thus $\{0\}$ is the only maximal regular ideal and $\mathsf{Spec}(L)$ is a singleton.

The Pierce representation of a ring

When $M \lhd S$ is a regular maximal ideal of the non-trivial ring S, the quotient ring S/M no longer has non-trivial idempotents. Consider the disjoint union of all these quotient rings and the projection to the spectrum which maps the whole of S/M to M:

$$p: \coprod_{M \in \mathsf{Spec}(S)} S/M \longrightarrow \mathsf{Spec}(S), \quad [s] \in S/M \mapsto M.$$

For each regular $I \lhd S$ and each element $s \in S$, we get a section of p on the open subset O_I:

$$\sigma_s^I : O_I \longrightarrow \coprod_{M \in \mathsf{Spec}(S)} S/M, \quad M \mapsto [s] \in S/M; \quad (p \circ \sigma_s^I)(N) = N.$$

We choose on $\coprod_{M \in \mathsf{Spec}(S)} S/M$ the final topology for all these local sections σ_s^I.

The Pierce representation theorem asserts that the continuous global sections of p

$$s: \mathsf{Spec}(S) \longrightarrow \coprod_{M \in \mathsf{Spec}(S)} S/M, \quad p \circ s = \mathsf{id}_{\mathsf{Spec}(S)},$$

provided with the pointwise addition and multiplication, constitute a ring isomorphic to S. All this is of course reminiscent of the theory of sheaves on a topological space and their description in terms of étale mappings.

Algebras over a ring

An S-algebra over a ring S is another ring A provided with a scalar multiplication by S:

$$S \times A \longrightarrow A, \quad (s, a) \mapsto sa.$$

The axioms of an S-algebra imply in particular that

$$\forall s \in S \ \forall a \in A \ \ sa = s(1a) = (s1)a$$

where thus $1 \in A$ and $s1 \in A$. Therefore, to define an S-algebra structure on a ring A, it suffices to define $s1$ for each $s \in S$. This reduces to giving a ring homomorphism

$$i_A : S \longrightarrow A, \quad s \mapsto s1.$$

Since an S-algebra is a ring, we can speak of $\mathsf{Spec}(A)$. The morphism i_A induces a continuous function

$$\mathsf{Spec}(i_A): \mathsf{Spec}(A) \longrightarrow \mathsf{Spec}(S)$$

because the image of an idempotent is an idempotent. This extends further as a contravariant functor

$$S_S: \mathsf{Alg}_S \longrightarrow \mathsf{Prof}/\mathsf{Spec}(S), \quad A \mapsto \Big(\mathsf{Spec}(i_A): \mathsf{Spec}(A) \to \mathsf{Spec}(S)\Big)$$

from the category of S-algebras to the category of profinite spaces over $\mathsf{Spec}(S)$.
This functor $\mathcal{S}_S$ has a right adjoint:

$$C_S \colon \mathsf{Prof}/\mathsf{Spec}(S) \longrightarrow \mathsf{Alg}_S,$$

$$C_S(X, \varphi) = \left\{ h \colon X \longrightarrow \coprod_{M \in \mathsf{Spec}(S)} S/M \,\middle|\, h \text{ continous, } p \circ h = \varphi \right\}$$

where the S-algebra structure on $C(X, \varphi)$ is defined pointwise.

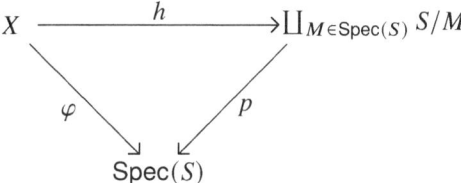

Split algebras over a ring

When $R \subseteq S$ is a ring extension, $S \otimes_R A$ is a S-algebra, for each R-algebra A. An R-algebra A is *split* by S when there exists a canonical isomorphism

$$C_S \mathcal{S}_S(S \otimes_R A) \cong S \otimes_R A.$$

This definition doesn't at all look like the corresponding definition in the case of fields, but it reduces nevertheless to the classical definition of split algebra in the case of fields. Indeed when $K \subseteq L$ is a field extension, we know that $\mathsf{Spec}(L)$ is a singleton. Therefore $\mathcal{S}_L$ is simply the functor

$$\mathsf{Spec} \colon \mathsf{Alg}_L \longrightarrow \mathsf{Prof}, \quad B \mapsto \mathsf{Spec}(B)$$

and the right adjoint to that functor is

$$C(-, L) \colon \mathsf{Prof} \longrightarrow \mathsf{Alg}_L, \quad X \mapsto C(X, L)$$

where L has the discrete topology and $C(X, L)$ indicates now the ring of continuous functions. Thus, a K-algebra A is split by L in the sense of ring theory when

$$C\Big(\mathsf{Spec}(L \otimes_K A), L\Big) \cong L \otimes_K A.$$

We shall prove that in the case of a finite-dimensional field extension $K \subseteq L$, a K-algebra A is split by L in the sense of field theory when there exists an isomorphism of L-algebras

$$L \otimes_K A \cong L^n,$$

where n is the dimension of A as a K-algebra. The field L admits 0 and 1 as its only idempotents, thus L^n admits as idempotent elements

$$(\varepsilon_i)_{i=1,\ldots,n}, \quad \varepsilon_i = 0, 1.$$

This implies that the spectrum of $L \otimes_K A \cong L^n$ is the discrete space with n elements. Therefore

$$C\Big(\mathrm{Spec}(L \otimes_K A), L\Big) \cong C(n, L) \cong L^n \cong L \otimes_K A$$

and we conclude that the K-algebra A is split by $K \subseteq L$ in the sense of the theory of rings. The same conclusion will hold in arbitrary dimension.

Galois extensions of rings

A ring extension $R \subseteq S$ is called a *Galois extension* when:

1. $R \subseteq S$ is a pure extension;
2. the R-algebra S is split by the extension $R \subseteq S$.

This implies that for each $(X, \varphi) \in \mathrm{Prof}/\mathrm{Spec}(S)$, the R-algebra $C_S(X, \varphi)$ is split by $R \subseteq S$.

When $K \subseteq L$ is a field extension, the only maximal regular ideal of L is $\{0\}$. In this case, the extension $K \subseteq L$, viewed as ring extension, is a Galois extension when $L = L/\{0\}$ is split by $K \subseteq L$: this is exactly the definition of a Galois extension of fields.

Internal categories and presheaves

A small category C can be seen as a set C_0 of objects, a set C_1 of arrows, and four operations d_0 (domain), d_1 (codomain), i (identity) and c (composition). Of course

$$d_0 \colon C_1 \longrightarrow C_0, \quad d_1 \colon C_1 \longrightarrow C_0, \quad i \colon C_0 \longrightarrow C_1.$$

The composition is defined on the set of successive pairs of arrows, that is, on the pullback of d_1 and d_0:

$$c \colon C_1 \times_{C_0} C_1 \longrightarrow C_1.$$

Of course, these data must satisfy the category axioms, which are easily expressed by the commutativity of some diagrams involving these arrows.

When $\mathcal{E}$ is a category with pullbacks, an internal category in $\mathcal{E}$ is defined in exactly the same way, choosing now C_0, C_1 to be objects in $\mathcal{E}$ and d_0, d_1, i, c, to be morphisms in $\mathcal{E}$. For example when $\mathcal{E}$ is the category of profinite topological spaces, we get the notion of profinite category.

Now, back to the set case, let us consider a functor $F \colon C \longrightarrow \mathrm{Set}$. Giving all the sets $F(C)$ for $C \in C_0$ is equivalent to giving the projection

$$p \colon [F] = \coprod_{C \in C_0} F(C) \longrightarrow C_0$$

which maps the whole of $F(C)$ to $C \in C_0$. Giving the action of F on the arrows then gives a mapping on the pullback of d_0 and p

$$C_1 \times_{C_0} [F] \longrightarrow [F]$$

which maps $\big(f\colon A \to B, a \in F(A)\big)$ to $F(f)(a)$. Here, again, the commutativity of diagrams translates the functoriality of F. An analogous description holds for contravariant functors.

Again, when $\mathcal{E}$ is a category with pullbacks and C is an internal category, an internal functor F from C to $\mathcal{E}$ is defined as an object $[F]$ of $\mathcal{E}$ defined over C_0, together with a morphism

$$\varphi\colon C_1 \times_{C_0} [F] \longrightarrow [F]$$

making commutative the *ad hoc* diagrams. In the case $\mathcal{E} = \mathsf{Set}$, $[F]$ is the disjoint union of all the sets $F(A)$, for all objects A of C; given a morphism $f\colon A \longrightarrow B$ and an element $a \in F(A)$, $\varphi(f, a) = F(f)(a)$.

The Galois groupoid of a ring extension

We should now define the Galois group of a Galois extension of rings … but this is no longer a group!

A groupoid is a category whose morphisms are all isomorphisms. A group can thus be seen as a groupoid with a single object, whose arrows are the elements of the group. In the case of rings, the objects of the Galois groupoid $\mathsf{Gal}[S : R]$ are the elements of $\mathsf{Spec}(S)$ … which is indeed a singleton when S is a field.

But clearly, since rings generalize the case of arbitrary dimension for fields, the Galois groupoid of a Galois extension $R \subseteq S$ of rings must be a profinite groupoid. Besides the profinite topology on the set $\mathsf{Spec}(S)$ of objects, we must thus have a profinite topology on the set of arrows, which makes all the category operations continuous: domain, codomain, identity, composition, inverse. Keeping in mind that the functor Spec is contravariant, the profinite Galois groupoid is defined as follows:

- $\mathsf{Spec}(S)$ is the space of objects;
- $\mathsf{Spec}(S \otimes_R S)$ is the space of arrows;
- the ring homomorphisms

$$S \rightrightarrows S \otimes_R S, \quad s \mapsto s \otimes 1,\ s \mapsto 1 \otimes s$$

  induce the domain and codomain operations

$$\mathsf{Spec}(S \otimes_R S) \rightrightarrows \mathsf{Spec}(S); \quad (f\colon M \to N) \mapsto M,\ N$$

- the ring homomorphism

$$S \otimes_R S \longrightarrow S, \quad s \otimes s' \mapsto ss'$$

  induces the identity operation

$$\mathsf{Spec}(S) \longrightarrow \mathsf{Spec}(S \otimes_R S), \quad M \mapsto \mathsf{id}_M;$$

- the ring homomorphism

$$S \otimes_R S \longrightarrow S \otimes_R S, \quad s \otimes s' \mapsto s' \otimes s$$

induces the inverse operation

$$\mathsf{Spec}(S \otimes_R S) \longrightarrow \mathsf{Spec}(S \otimes_R S), \quad f \mapsto f^{-1};$$

- finally one observes that $\mathsf{Spec}(S \otimes_R S \otimes_R S)$ is homeomorphic to the subspace of pairs of composable morphisms; the ring homomorphism

$$S \otimes_R S \longrightarrow S \otimes_R S \otimes_R S, \quad s \otimes s' \mapsto s \otimes 1 \otimes s'$$

then induces the composition operation

$$\mathsf{Spec}(S \otimes_R S) \times_{\mathsf{Spec}(S)} \mathsf{Spec}(S \otimes_R S) \longrightarrow \mathsf{Spec}(S \otimes_R S), \quad (f, g) \mapsto f \circ g.$$

We shall keep writing $\mathsf{Gal}[S : R]$ for this profinite Galois groupoid.

In the case of a finite-dimensional Galois extension $K \subseteq L$ of fields, we know that

$$L \otimes_K L \cong L^n$$

because L is a K-algebra split by L. We know also that

$$\mathsf{Spec}(L \otimes_K L) = \mathsf{Spec}(L^n) = n$$

is a discrete space with n elements and that $\mathsf{Spec}(L)$ is a singleton. This implies that the Galois groupoid of the Galois extension $K \subseteq L$, seen as a Galois extension of rings, is a discrete group with n elements. It is easily observed that this group is precisely the Galois group $\mathsf{Gal}[L : K]$ of the field extension $K \subseteq L$. The case of arbitrary dimension follows easily.

The action of the Galois groupoid

Finally we have to generalize the notion of "set provided with an action of the Galois group".

If G is a group, an action of G on a set X is a mapping

$$G \times X \longrightarrow X, \quad (g, x) \mapsto gx$$

or equivalently, a mapping

$$X \longrightarrow X, \quad x \mapsto gx$$

for each element $g \in G$. When we view G as a groupoid $\mathcal{G}$ with a single object, the elements of G are the arrows of the groupoid $\mathcal{G}$ and a G-set reduces to giving a contravariant functor

$$\mathcal{G} \longrightarrow \mathsf{Set}, \quad * \mapsto X, \quad g \mapsto (x \mapsto gx)$$

from the groupoid $\mathcal{G}$ to the category Set of sets.

The generalization is easy: when $\mathcal{G}$ is a groupoid, it suffices to consider the contravariant functors from $\mathcal{G}$ to Set. Such functors are called the *presheaves* on $\mathcal{G}$.

However, in the case of the profinite groupoid $\mathrm{Gal}[S : R]$ of a Galois ring extension $R \subseteq S$, we must of course consider "profinite presheaves": that is, internal contravariant functors in the category of profinite spaces. We thus consider internal contravariant functors

$$p\colon [P] \longrightarrow \mathrm{Spec}(S)$$

from the profinite groupoid $\mathrm{Gal}[S : R]$ to the category Prof of profinite topological spaces.

This concludes the list of the necessary ingredients for stating the Galois theorem for rings.

The Galois theorem for rings

The Galois theorem for rings now takes the following form:

If $R \subseteq S$ is a Galois extension of rings, the category of R-algebras split by S is equivalent to the category of internal profinite presheaves over the internal profinite Galois groupoid $\mathrm{Gal}[S : R]$.

This theorem contains as particular cases the Galois theorems for Galois extensions of fields, in finite or arbitrary dimension.

Part I
Some Galois Theorems for Fields

Chapter 1
The Classical Galois Theorem

Convention. *In this chapter, and the whole book, all fields are commutative.*

Abstract This first chapter presents the classical Galois theorem for fields. A finite-dimensional extension of fields $K \subseteq L$ is a *Galois extension* when every element $l \in L$ is root of a polynomial $p(X) \in K[X]$ which factors in $L[X]$ into distinct linear factors. The *Galois group* $\mathsf{Gal}[L : K]$ of that extension is the group of all field endomorphisms (and thus automorphisms) of L which fix all the elements of K. The Galois theorem exhibits a bijection between the subgroups of the Galois group and the intermediate field extensions $K \subseteq M \subseteq L$.

1.1 Algebraic Extensions of Fields

Let us first recall two elementary results.

Lemma 1.1 *Each field homomorphism $f : K \longrightarrow L$ is injective.*

Proof If $0 \neq k \in K$, $f(k) \cdot f(k^{-1}) = f(1) = 1$, thus $f(k) \neq 0$. $\qquad\square$

In the situation of Lemma 1.1, we shall generally refer to the field extension $K \subseteq L$ without mentioning the field homomorphism f, which we thus think of as a canonical inclusion. The multiplication of L, restricted to $K \times L$, presents L as a K-vector space; we shall write $\dim [L : K]$ to indicate its dimension.

Lemma 1.2 *Let $K \subseteq M \subseteq L$ be field extensions. Then*

$$\dim [L : K] = \dim [L : M] \cdot \dim [M : K].$$

Proof If $(l_i)_{i \in I}$ is a base of L over M and $(m_j)_{j \in J}$ is a base of M over K, the products $(l_i \cdot m_j)_{(i,j) \in I \times J}$ constitute a base of L over K. $\qquad\square$

Definition 1.3 A field extension $K \subseteq L$ is *algebraic* when each element $l \in L$ is a root in L of a polynomial $p(X) \in K$.

F. Borceux, *Galois Theories of Fields and Rings*, Coimbra Mathematical Texts 2,
https://doi.org/10.1007/978-3-031-58460-2_2

Proposition 1.4 *Consider field extensions $K \subseteq M \subseteq L$. If $K \subseteq L$ is algebraic, then $K \subseteq M$ and $M \subseteq L$ are algebraic as well.*

Proof Each element $l \in L$, thus also each element $l \in M$, is a root of a polynomial $p(X) \in K[X] \subseteq M[X]$. $\qquad\square$

Proposition 1.5 *Each finite-dimensional field extension $K \subseteq L$ is algebraic.*

Proof For each element $l \in L$, the sequence of $n + 1$ elements $1, l, l^2, l^3, \ldots, l^n$, with $n = \dim[L : K]$, cannot be free over K, thus there exists a linear combination

$$a_n l^n + a_{n-1} l^{n-1} + \cdots + a_2 l^2 + a_1 l + a_0 = 0, \quad a_i \in K.$$

It suffices to choose

$$p(X) = a_n X^n + a_{n-1} X^{n-1} + \cdots + a_2 X^2 + a_1 X + a_0 \in K[X]$$

to get $p(l) = 0$. $\qquad\square$

Proposition 1.6 *Let $K \subseteq L$ be an algebraic extension of fields. For each element $l \in L$, there exists a unique polynomial $p(X) \in K[X]$ such that:*

1. *$p(l) = 0$;*
2. *the coefficient of the term of maximal degree of $p(X)$ is 1;*
3. *$p(X)$ divides in $K[X]$ each polynomial $q(X) \in K[X]$ such that $q(l) = 0$.*

In particular,

4. *$p(X)$ has the minimal degree among all the polynomials $0 \neq q(X) \in K[X]$ such that $q(l) = 0$;*
5. *$p(X)$ is irreducible in $K[X]$.*

This polynomial $p(X)$ is called the minimal polynomial *of l.*

Proof From Definition 1.3, there exists a polynomial $p(X)$ such that $p(l) = 0$; with no loss of generality we choose it of minimal non-zero degree with a leading coefficient equal to 1. This polynomial is irreducible in $K[X]$. Indeed if $p(X) = \alpha(X)\beta(X)$ with $\alpha(X)$ and $\beta(X)$ non-constant, then $p(l) = 0$ implies that $\alpha(l) = 0$ or $\beta(l) = 0$, which contradicts the minimality of the degree of $p(X)$.

Next if $q(X) \in K[X]$ is such that $q(l) = 0$, let us divide $q(X)$ by $p(X)$ in $K[X]$:

$$q(X) = \gamma(X)p(X) + \delta(X), \quad \text{degree } \delta(X) < \text{degree } p(X).$$

We get $\beta(l) = 0$ because $p(l) = 0$ and $q(l) = 0$. Thus $\beta(X) = 0$, by minimality of the degree of $p(X)$.

Further, if $p_1(X)$, $p_2(X)$ are two polynomials satisfying conditions 1-2-3, then each of them divides the other. So each must be a scalar multiple of the other one and since they have the same leading coefficient 1, they are equal. $\qquad\square$

Proposition 1.7 *Let $K \subseteq L$ be an algebraic extension of fields and $l \in L$ an element with minimal polynomial $p(X) \in K[X]$.*

1. *The subfield $K(l) \subseteq L$ generated by K and l is isomorphic to the quotient of $K[X]$ by the principal ideal generated by $p(X)$.*
2. $\dim \left[K(l) : K\right]$ *is equal to the degree of $p(X)$.*

Proof Let

$$p(X) = X^n + a_{n-1}X^{n-1} + \cdots + a_1X + a_0$$

be the minimal polynomial of $l \in L$. The quotient of the ring $K[X]$ by the principal ideal generated by $p(X)$ can equivalently be described as the set of polynomials

$$K_p[X] = \left\{k_{n-1}X^{n-1} + \cdots + k_1X + k_0 \middle| k_i \in K\right\}$$

of degree at most $n - 1$, with the operations defined modulo the relation

$$X^n = -a_{n-1}X^{n-1} - \cdots - a_1X - a_0.$$

This is both a ring and a K-vector space[1] of dimension n, the degree of $p(X)$, since it admits the basis $1, X, \ldots, X^{n-1}$.

In the same way, since $p(l) = 0$, the relation

$$l^n = -a_{n-1}l^{n-1} - \cdots - a_1l - a_0$$

implies that

$$K(l) = \left\{k_{n-1}l^{n-1} + \cdots + k_0 \middle| k_i \in K\right\}$$

is a subring and sub-K-vector space[2] of L, generated by K and l. As a K-vector space, it is generated by $1, l, \ldots, l^{n-1}$, thus is finite-dimensional.

Choose $0 \neq a \in K(l)$. Multiplying by a

$$K(l) \longrightarrow K(l), \quad b \mapsto ab$$

is a K-linear mapping. This mapping is injective, because a admits the inverse a^{-1} in L. Thus this mapping is a K-linear isomorphism, because the K-vector space $K(l)$ is finite-dimensional. Thus, there exists a $b \in K(l)$ such that $ab = 1 \in K(l)$ and therefore $K(l)$ is a field.

To conclude, let us consider the ring homomorphism

$$\gamma: K_p[X] \longrightarrow K(l), \quad q(X) \mapsto q(l).$$

This is an injective K-linear mapping: indeed, $q(l) = 0$ implies $q(X) = 0$, because the degree of $q(X)$ is strictly less than the degree of the minimal polynomial $p(X)$ of l (see Proposition 1.6). But by definition of $K(l)$, γ is trivially surjective, thus is an isomorphism. $\qquad\qquad\square$

---

[1] In fact, a K-algebra; see Definition 2.1.

[2] Again, a sub-K-algebra.

Definition 1.8 Let $K \subseteq L$ be a field extension. A field endomorphism of L which fixes all the elements of K

$$f: L \longrightarrow L, \quad \forall k \in K \ \ f(k) = k,$$

is called a *K-endomorphism* of L.

The K-automorphisms are closely related to the permutations of the roots of the minimal polynomials.

Proposition 1.9 *Let $K \subseteq L$ be a field extension. Each K-endomorphism of L induces a permutation of the roots of the minimal polynomial of every element $l \in L$.*

Proof Let $l \in L$ have the minimal polynomial $p(X) \in K[X]$ and f, a K-endomorphism of L. Since f fixes all the elements of K, for each root $l' \in L$ of $p(X)$

$$p\big(f(l')\big) = f\big(p(l')\big) = f(0) = 0$$

and thus $f(l')$ is still a root of $p(X)$. Thus f induces a mapping

$$\hat{f} \colon \{l' \in L | p(l') = 0\} \longrightarrow \{l" \in L | p(l") = 0\}, \quad l' \mapsto f(l').$$

The set of roots of $p(X)$ is finite and f is injective (see Lemma 1.1); thus $\hat{f}$ is a bijection. $\square$

Proposition 1.10 *Let $K \subseteq L$ be an algebraic extension of fields. Each K-endomorphism of L is an automorphism.*

Proof Each K-endomorphism $f: L \longrightarrow L$ is injective (see Lemma 1.1). It remains to prove that f is surjective. Let us choose an element $l \in L$ with minimal polynomial $p(X)$. By Proposition 1.9, there exists a root $l' \in L$ of $p(X)$ such that $f(l') = l$. $\square$

Definition 1.11 Let $K \subseteq L$ be an algebraic field extension. The group $\mathsf{Gal}[L : K]$ of K-automorphisms of L is called the *Galois group* of the extension $K \subseteq L$.

1.2 Galois Extensions of Fields

Definition 1.12 Call a field extension $K \subseteq L$ a *Galois extension* when each element $l \in L$ is root of a polynomial $q(X) \in K[X]$ which admits in $L[X]$ a factorization in distinct linear factors.[3]

Clearly, "distinct linear factors" means "distinct up to a scalar multiple": two linear factors as $X - a$ and $l(X - a)$, $0 \neq l \in L$, are not considered to be distinct.

---

[3] Let us mention two related weaker notions. An algebraic extensions $K \subseteq L$ is *separable* when the minimal polynomial of each element $l \in L$ has only simple roots in L; it is *normal* when every irreducible polynomial in $K[X]$ which has a root in L splits into linear factors in $L[X]$; separable and normal is equivalent to Galois.

This definition admits an equivalent formulation:

Proposition 1.13 *A field extension $K \subseteq L$ is Galois when*

1. *the extension $K \subseteq L$ is algebraic;*
2. *the minimal polynomial $p(X) \in K[X]$ of each element $l \in L$ admits a decomposition into distinct linear factors in $L[X]$.*

Proof Clearly, the conditions in Proposition 1.13 implies those in Definition 1.12. Conversely, and with the notation of these two statements, $p(X)$ is a factor of $q(X)$ (see Proposition 1.6), thus admits a decomposition into distinct linear factors in $L[X]$.   □

Proposition 1.14 *Consider field extensions $K \subseteq M \subseteq L$. If $K \subseteq L$ is Galois, then also $M \subseteq L$ is a Galois extension.*

Proof Each element $l \in L$ is a root of a polynomial $p(X) \subseteq K[X] \subseteq M[X]$ which admits a decomposition into linear factors in $L[X]$.   □

Lemma 1.15 *Consider field extensions $K \subseteq N \subseteq L$, with $K \subseteq L$ Galois. If $l \in L$, $l \notin N$, and $f: N \longrightarrow L$ is a field homomorphism, then f extends as a field homomorphism $\tilde{f}: N(l) \longrightarrow L$.*

Proof The homomorphism f, applied to the coefficients of a polynomial, induces a ring homomorphism

$$\overline{f}: N[X] \longrightarrow L[X].$$

By Proposition 1.14, we know that $N \subseteq L$ is a Galois extension. By Proposition 1.6, the minimal polynomial $q(X) \in N[X]$ of l over N divides in $N[X]$ the minimal polynomial $p(X) \in K[X]$ of l over K. Thus $\overline{f}(q(X))$ divides $\overline{f}(p(X))$ in $L[X]$. But $\overline{f}(p(X)) = p(X)$ because f fixes the elements of K. The polynomial $p(X)$ admits a decomposition into distinct linear factors over L, thus the same holds for its factor $\overline{f}(q(X))$. Therefore the polynomial $\overline{f}(q(X))$ admits at least one root $l' \in L$.

Clearly, $X - l'$ is the minimal polynomial of $l' \in L$ over L. Thus, using Proposition 1.7 twice, we obtain an extension $\tilde{f}$ of f, where $\langle \rangle$ indicates the "ideal generated by":

$$\tilde{f}: N(l) \cong \frac{N[X]}{\langle q(X) \rangle} \xrightarrow{\ \overline{f}\ } \frac{L[X]}{\langle \overline{f}(q(X)) \rangle} \overset{p}{\twoheadrightarrow} \frac{L[X]}{\langle X - l' \rangle} \cong L(l') \cong L.$$

The quotient p exists because $X - l'$ is a factor of $\overline{f}(q(X))$, thus $\langle \overline{f}(q(X)) \rangle \subseteq \langle X - l' \rangle$.   □

Proposition 1.16 *Consider field extensions $K \subseteq M \subseteq L$ with $K \subseteq L$ a finite-dimensional Galois extension. Each K-automorphism of M extends as a K-automorphism of L.*

Proof Let $f: M \longrightarrow M$ be a K-homomorphism. By Proposition 1.10, it suffices to construct an extension $\tilde{f}: L \longrightarrow L$ of f: it will automatically be a K-automorphism.

Let us use Lemma 1.15 inductively on the natural numbers to construct a sequence $f_n \colon N_n \longrightarrow L$ of extensions of f.

- $N_0 = M$, $f_0 = f \colon N_0 \longrightarrow L$.
- If $N_n = L$, $f_n \colon L \longrightarrow L$ is an extension of f and the proposition is proved. If not, choose $l \in L$, $l \notin N_n$; put $N_{n+1} = N_n(l)$ and $f_{n+1} = \tilde{f}_n$, where $\tilde{f}_n$ is an extension given by Lemma 1.15.

Since $\dim [L : M]$ is finite, the process will stop at some $N_n = L$, after finitely many steps. $\qquad\qquad\qquad\square$

Proposition 1.17 *Consider field extensions $K \subseteq M \subseteq L$ with $K \subseteq M$ a Galois extension. Each K-endomorphism of L admits a restriction $f|_M \colon M \longrightarrow M$.*

Proof Consider $l \in M$ with minimal polynomial $p(X) \in K[X]$. Since f is a K-homomorphism

$$p\big(f(l)\big) = f\big(p(l)\big) = f(0) = 0.$$

Thus $f(l)$ is a root of $p(X)$ and, since $K \subseteq M$ is a Galois extension, this root lies in M. $\qquad\qquad\qquad\square$

Definition 1.18 Let $K \subseteq L$ be an algebraic extension of fields. Two elements $l_1, l_2 \in L$ are *conjugate over K* when they have the same minimal polynomial.

Proposition 1.19 *Consider a field extension $K \subseteq L$ and $p(X) \in K[X]$, the minimal polynomial of $l \in L$. Each root $l' \in L$ of $p(X)$ is conjugate to l over K.*

Proof The minimal polynomial $q(X) \in K[X]$ of l' divides $p(X)$ (see Proposition 1.6). Since $p(X)$ is irreducible, $q(X) = p(X)$. $\qquad\qquad\qquad\square$

The following result exhibits the link between K-automorphisms and the permutation of roots, in which Galois was interested (see the introduction).

Proposition 1.20 *Let $K \subseteq L$ be a finite-dimensional Galois extension of fields and $l_1, l_2 \in L$. The following conditions are equivalent:*

1. *the elements l_1, l_2 are conjugate over K;*
2. *there exists a K-automorphism $f \colon L \longrightarrow L$ such that $f(l_1) = l_2$.*

Proof If l_1, l_2 are conjugate with minimal polynomial $p(X) \in K[X]$, a double application of Proposition 1.7 yields a K-isomorphism

$$g \colon K(l_1) \cong \frac{K[X]}{\langle p(X) \rangle} \cong K(l_2)$$

such that $g(l_1) = l_2$. By Proposition 1.16, we can extend g as a K-automorphism of L.

Conversely if $p(X) \in K[X]$ is the minimal polynomial of l_1, we obtain

$$p(l_2) = p\big(f(l_1)\big) = f\big(p(l_1)\big) = f(0) = 0.$$

because f is a K-homomorphism. Thus l_2 is conjugate to l_1 over K (see Proposition 1.19). $\qquad\qquad\qquad\square$

1.3 The Classical Galois Theorem for Fields

If $K \subseteq L$ is a field extension and $G \subseteq \mathrm{Gal}[L : K]$ is a subgroup of its Galois group, we write

$$\mathrm{Fix}(G) = \{l \in L | \forall f \in G \ f(l) = l\}$$

for the set of elements of L fixed by all the automorphisms in G.

Lemma 1.21 *Given a field extension $K \subseteq L$ and a subgroup $G \subseteq \mathrm{Gal}[L : K]$, we obtain field extensions*

$$K \subseteq \mathrm{Fix}(G) \subseteq L.$$

Proof Each $f \in G$ is a field homomorphism fixing the elements of K. □

Definition 1.22 A *Galois connection* between two partially ordered sets $(A, \leq)$ and $(B, \leq)$ is a pair of monotone mappings

$$(A, \leq) \xrightarrow[\quad f \quad]{\quad g \quad} (B, \leq)$$

such that, for all elements $a \in A$ and $b \in B$,

$$a \leq g(b) \text{ iff } b \leq f(a).$$

Notice that a partially ordered set $(A, \leq)$ can be seen as a category, whose objects are the elements of A and a single morphism is put from a to a' when $a \leq a'$. In this spirit, a Galois connection is a pair of adjoint functors, in the sense of Definition 5.1.

Proposition 1.23 *Let $K \subseteq L$ be an algebraic field extension. The functors*

$$\{M \mid K \subseteq M \subseteq L\} \xrightarrow[\quad \mathrm{Fix} \quad]{\quad \mathrm{Gal}[L : -] \quad} \{G \mid G \subseteq \mathrm{Gal}[L : M]\},$$

where M is a subfield and G a subgroup, constitute a Galois connection.

Proof The statement means that

$$M \subseteq M' \Rightarrow \mathrm{Gal}[L : M'] \subseteq \mathrm{Gal}[L : M],$$
$$G \subseteq G' \Rightarrow \mathrm{Fix}(G') \subseteq \mathrm{Fix}(G),$$
$$M \subseteq \mathrm{Fix}(G) \Leftrightarrow G \subseteq \mathrm{Gal}[L : M].$$

The first two implications are obvious; the last one is a direct consequence of

$$M \subseteq \mathrm{Fix}(\mathrm{Gal}[L : M]), \quad G \subseteq \mathrm{Gal}\big[L : \mathrm{Fix}(G)\big]. \qquad \Box$$

We are ready to state the classical Galois theorem:

Theorem 1.24 (Galois Theorem) *Let $K \subseteq L$ be a finite-dimensional Galois extension of fields. In that case, the Galois connection*

$$\{M \mid K \subseteq M \subseteq L\} \underset{\mathsf{Fix}}{\overset{\mathsf{Gal}[L : -]}{\rightleftarrows}} \{G \mid G \subseteq \mathsf{Gal}[L : M]\}$$

is an isomorphism and, for each M,

$$\dim [L : M] = \#\mathsf{Gal}[L : M],$$

where # indicates the cardinal of the set.

Proof There exist many proofs of this Galois theorem: short and elegant proofs using involved results of field theory, and long and technical proofs using only elementary properties. Since this book intends to present Galois theory to readers who have not necessarily specialized in field theory, our choice is that of an "elementary proof".

First, let us prove by induction on $\dim [L : M]$ that $\dim [L : M] = \#\mathsf{Gal}[L : M]$. If $\dim [L : M] = 1$, then $M = L$ and the only L-automorphism of L is the identity on L.

Let us now assume that $\dim [L' : M'] = \#\mathsf{Gal}[L' : M']$ for each $K \subseteq M' \subseteq L'$ with $\dim [L' : M'] < n$. We consider $K \subseteq M \subseteq L$ where $\dim [L : M] = n$. If $l \in L$, $l \notin M$, admits $p(X) \in M[X]$ as minimal polynomial of degree r, let us consider the extension $M \subseteq M(l) \subseteq L$. Since $l \notin M$, we have $M \neq M(l)$ and therefore $\dim [L : M(l)] < n$. But

$$\dim [M(l) : M] \cdot \dim [L : M(l)] = \dim [L : M];$$

thus $\dim [L : M(l)] = \frac{n}{r}$. The induction assumption implies that there exist exactly $\frac{n}{r}$ $M(l)$-automorphisms

$$f_1, \ldots, f_{\frac{n}{r}} : L \longrightarrow L.$$

Let us write $l_1, \ldots, l_r$ for the roots of $p(X)$ in L. For each index j, we get an M-automorphism $g_j : L \longrightarrow L$ such that $g_j(l) = l_j$ (see Proposition 1.20). Let us define

$$h_{ij} : L \longrightarrow L, \quad h_{ij} = g_j \circ f_i.$$

In this way we obtain n automorphisms h_{ij} of L. We must still prove that these are exactly the elements of the Galois group $\mathsf{Gal}[L : M]$.

Let us observe that these h_{ij} are M-automorphisms, thus elements of $\mathsf{Gal}[L : M]$. Moreover these elements are distinct, because

$$\begin{aligned}
h_{ij} = h_{i'j'} &\Rightarrow g_j f_i(l) = g_{j'} f_{i'}(l) \\
&\Rightarrow g_j(l) = g_{j'}(l) && (f_i, f_j \text{ fix } M(l)) \\
&\Rightarrow l_j = l_{j'} && (\text{definition of } g_j, g_{j'}) \\
&\Rightarrow j = j' && (M \subseteq L \text{ Galois extension}) \\
&\Rightarrow g_j = g_{j'}.
\end{aligned}$$

Next

$$h_{ij} = h_{i'j} \Rightarrow g_j f_i = g_j f_{i'} \Rightarrow f_i = f_{i'}$$

because g_j is injective.

Finally, if $f \colon L \longrightarrow L$ is an M-automorphism, it is necessarily one of the h_{ij}. Indeed, $p(f(l)) = f(p(l)) = f(0) = 0$, thus $f(l) = l_j$ for some index j. Therefore $(g_j^{-1} \circ f)(l) = g_j^{-1}(l_j) = l$ and $g_j^{-1} \circ f$ is an M-automorphism which fixes l. Thus $g_j^{-1} \circ f$ is a $M(l)$-automorphism and $g_j^{-1} \circ f = f_i$ for some index i. So, $f = g_j \circ f_i = h_{ij}$.

Next, let us prove the formula $\dim \big[L : \mathsf{Fix}(G)\big] = \#G$ for each subgroup $G \subseteq \mathsf{Gal}[L : K]$. It suffices to prove $\dim \big[L : \mathsf{Fix}(G)\big] \le \#G$ because, using the first part of the proof and Proposition 1.23, we get

$$\#G \le \#\mathsf{Gal}\big[L : \mathsf{Fix}(G)\big] = \dim \big[L : \mathsf{Fix}(G)\big] \le \#G.$$

Still using the first part of the proof,

$$\dim [L : K] = \#\mathsf{Gal}[L : K].$$

Thus $\mathsf{Gal}[L : K]$ and therefore G are finite. Put $\#G = n$; we must prove that $\dim \big[L : \mathsf{Fix}(G)\big] \le n$.

We develop the proof by a reduction *ad absurdum*. Let us suppose the existence of elements $l_1, \ldots, l_{n+1}$ in L, linearly independent over $\mathsf{Fix}(G)$. Let us write $g_1, \ldots, g_n$ for the n elements of G. Let us further consider the system of homogeneous equations

$$\left. \begin{aligned} g_1(l_1)X_1 + \cdots + g_1(l_{n+1})X_{n+1} &= 0, \\ &\vdots \\ g_n(l_1)X_1 + \cdots + g_n(l_{n+1})X_{n+1} &= 0. \end{aligned} \right\} \tag{1}$$

Since the number of unknowns is greater than the number of equations, there exists a non-zero solution. Let us choose such a solution with the minimal number of non-zero components. Up to renumbering the unknowns, assume that this solution has the form $(\alpha_0, \ldots, \alpha_r, 0, \ldots, 0)$, with each α_i non-zero. We obtain a system

$$\left. \begin{aligned} g_1(l_1)X_1 + \cdots + g_1(l_r)X_r &= 0, \\ &\vdots \\ g_n(l_1)X_1 + \cdots + g_n(l_r)X_r &= 0 \end{aligned} \right\} \tag{2}$$

admitting a solution whose components are all non-zero. Let us fix $g \in G$, apply g to all these equations and compute the values at $\alpha_1, \ldots, \alpha_r$:

$$\left. \begin{aligned} gg_1(l_1)g(\alpha_1) + \cdots + gg_1(l_r)g(\alpha_r) &= 0, \\ &\vdots \\ gg_n(l_1)g(\alpha_1) + \cdots + gg_n(l_r)g(\alpha_r) &= 0. \end{aligned} \right\} \tag{3}$$

The elements gg_i constitute a permutation of the elements of G; thus up to a permutation of the order of the equations, we can re-write our system (3) as

$$\left.\begin{aligned}g_1(l_1)g(\alpha_1) + \cdots + g_1(l_r)g(\alpha_r) &= 0, \\ &\vdots \\ g_n(l_1)g(\alpha_1) + \cdots + g_n(l_r)g(\alpha_r) &= 0.\end{aligned}\right\} \tag{4}$$

Let us now multiply our system (4) by α_r and our system (2), computed at α_i, by $g(\alpha_r)$. Subtracting these two systems yields the following system:

$$\left.\begin{aligned}g_1(l_1)\big(\alpha_r g(\alpha_1) - \alpha_1 g(\alpha_r)\big) + \cdots \\ + g_1(l_{r-1})\big(\alpha_r g(\alpha_{r-1}) - \alpha_{r-1}g(\alpha_r)\big) &= 0, \\ &\vdots \\ g_n(l_1)\big(\alpha_r g(\alpha_1) - \alpha_1 g(\alpha_r)\big) + \cdots \\ + g_n(l_{r-1})\big(\alpha_r g(\alpha_{r-1}) - \alpha_{r-1}g(\alpha_r)\big) &= 0.\end{aligned}\right\} \tag{5}$$

The system (5) presents a solution of the system (1) with $r - 1$ non-zero components. The minimal property of the solution $(\alpha_0, \ldots, \alpha_r, 0, \ldots, 0)$ implies that the solution of system (5) is a zero solution. Thus $\alpha_r g(\alpha_i) = \alpha_i g(\alpha_r)$ for all indices $i \le r - 1$. This implies

$$\alpha_i \alpha_r^{-1} = g(\alpha_i)g(\alpha_r)^{-1} = g(\alpha_i)g(\alpha_r^{-1}) = g(\alpha_i \alpha_r^{-1}).$$

Since $g \in G$ is arbitrary, we obtain $\alpha_i \alpha_r^{-1} \in \mathsf{Fix}(G)$. Putting $m_i = \alpha_i \alpha_r^{-1} \in \mathsf{Fix}(G)$, we get $\alpha_i = m_i \alpha_r$ with $i < r$ and $m_i \in \mathsf{Fix}(G)$. Putting further $m_r = 1 \in \mathsf{Fix}(G)$, we get this time $\alpha_i = m_i \alpha_r$ for all indices $i \le r$. The first equation of system (2), computed on the elements α_i, gives

$$\begin{aligned}0 &= g_1(l_1)\alpha_1 + \cdots + g_1(l_r)\alpha_r \\ &= g_1(l_1)m_1\alpha_1 + \cdots + g_1(l_r)m_r\alpha_r \\ &= \alpha_r\big(g_1(l_1)m_1 + \cdots + g_1(l_r)m_r\big) \\ &= \alpha_r\big(g_1(l_1)g_1(m_1) + \cdots + g_1(l_r)g_1(m_r)\big) \text{ because } m_i \in \mathsf{Fix}(G) \\ &= \alpha_r g_1(l_1 m_1 + \cdots + l_r m_r).\end{aligned}$$

We know that $\alpha_r \ne 0$ and g_1 is injective; thus $l_1 m_1 + \cdots + l_r m_r = 0$. This contradicts the linear independence of the elements l_i over $\mathsf{Fix}(G)$ and concludes the proof of the formula $\dim\big[L : \mathsf{Fix}(G)\big] = \#G$.

The Galois theorem now follows easily. Beginning with $K \subseteq M \subseteq L$, we have $M \subseteq \mathsf{Fix}\big(\mathsf{Gal}[L : M]\big)$ by Proposition 1.23. We must prove that when $l \in L, l \notin M$, there exists an $f : L \longrightarrow L$ which fixes the elements of M, but not l. If $p(X)$ is the minimal polynomial of l over M, $p(X)$ does not have degree 1 because $l \notin M$. Thus $p(X)$ has at least two distinct roots in L, because $M \subseteq L$ is a Galois extension (see Proposition 1.14). If $l' \ne l$ is another root of $p(X)$, Proposition 1.20 implies the existence of an M-automorphism f of L such that $f(l) = l'$.

Now let us choose $G \subseteq \mathsf{Gal}[L : K]$. Proposition 1.23 implies

$$G \subseteq \mathsf{Gal}\big[L : \mathsf{Fix}(G)\big].$$

The anterior parts of this proof show that

$$\#G \leq \#\mathsf{Gal}\big[L : \mathsf{Fix}(G)\big] = \dim\big[L : \mathsf{Fix}(G)\big] = \#G.$$

Since both cardinals are finite, $G = \mathsf{Gal}\big[L : \mathsf{Fix}(G)\big]$. □

Chapter 2
The Galois Theorem of Grothendieck

Convention. *In this chapter, all fields are commutative and all rings and algebras are commutative with unit.*

Abstract The Grothendieck approach to the Galois theorem extends the classical Galois bijection to an equivalence of categories. If $K \subseteq L$ is a finite-dimensional Galois extension of fields, a finite-dimensional K-algebra A is split by L when each element $a \in A$ is a root of a polynomial $p(X) \in K[X]$ which factors in $L[X]$ into distinct linear factors. The corresponding Galois theorem exhibits a contravariant equivalence between the category of finite-dimensional K-algebras split by L and the category of finite sets provided with an action of the Galois group $\mathsf{Gal}[L : K]$. This contains the classical Galois theorem. Indeed, via this equivalence, the K-algebras $K \subseteq M \subseteq L$ split by L coincide with the intermediate field extensions and are in bijection with the quotients of $\mathsf{Gal}[L : K]$, which is finite and viewed here as acting on itself. It is a classical result of the theory of group actions that these quotients are themselves in bijection with the subgroups of $\mathsf{Gal}[L : K]$.

2.1 Algebras Over a Field

This section recalls some basic results of the theory of algebras over a field.

Definition 2.1 Let K be a field. A K-algebra is a quadruple $(A, +, \times, \cdot)$ where

1. $(A, +, \times)$ is a ring (commutative, with unit);
2. $(A, +, \cdot)$ is a K-vector space;
3. $\forall k \in K \; \forall a, a' \in A \;\; k \cdot (a \times a') = (k \cdot a) \times a'$.

We write Alg_K for the category of K-algebras.

In general, we shall just write $k \cdot a = ka$ and $a \times a' = aa'$. The ring $K[X]$ of polynomials is obviously a K-algebra, as well as each power K^n, with the componentwise operations. An ideal $I \triangleleft A$ of a K-algebra is just an ideal I of the ring $(A, +, \times)$; in particular, I is a sub-K-vector space of A:

F. Borceux, *Galois Theories of Fields and Rings*, Coimbra Mathematical Texts 2,
https://doi.org/10.1007/978-3-031-58460-2_3

$$\forall k \in K \ \forall a \in A \quad k \cdot a = k \cdot (1 \times a) = (k \cdot 1) \times a$$

where $1 \in A$ is the unit of the ring multiplication. The quotient A/I is thus still a K-algebra.

Proposition 2.2 *Let K be a field and A a K-algebra. The following conditions are equivalent:*

1. *A is a field;*
2. *the only ideals of A are (0) and A.*

Under these conditions,

$$K \longrightarrow A, \quad k \mapsto k1$$

is a field extension.

Proof Let A be a field and $I \triangleleft A$ an ideal containing $0 \neq i \in I$. Since $1 = i^{-1}i \in I$, we have $a = a1 \in I$ for all $a \in A$. Conversely if $0 \neq a \in A$, then aA is a non-zero ideal of A, thus $aA = A$; in particular, there exists a $b \in A$ such that $ab = 1$. Under these conditions

$$K \longrightarrow A, \quad k \mapsto k1$$

is a field homomorphism and therefore, is injective (see Lemma 1.1).   □

Corollary 2.3 *Let K be a field and $f : A \longrightarrow B$ a surjective homomorphism of K-algebras. When B is a field, the kernel of f is a maximal ideal of A.*

Proof We have $B \cong A/\mathsf{Ker}\, f$. If $\mathsf{Ker}\, f \subseteq I$ with I a proper ideal, let us consider the quotient $q : A/\mathsf{Ker}\, f \longrightarrow A/I$. The kernel of q is an ideal of the field $B = A/\mathsf{Ker}\, f$, thus is trivial (see Proposition 2.2). Since I is a proper ideal, $A/I \neq (0)$; thus since q is surjective, $\mathsf{Ker}\, q \neq A/\mathsf{Ker}\, f$. Therefore $\mathsf{Ker}\, q = [0]$ and q is injective; so q is an isomorphism and $\mathsf{Ker}\, f = I$. This proves that $\mathsf{Ker}\, f$ is maximal.   □

Proposition 2.4 *Let K be a field. Each ideal of the K-algebra $K[X]$ is principal.*

Proof Let $I \subseteq K[X]$ be a non-zero ideal and $p(X) \in I$ a non-zero polynomial of minimal degree in I. For each polynomial $q(X) \in I$, let us consider its division by $p(X)$:

$$q(X) = p(X) \cdot \alpha(X) + \beta(X), \quad \text{degree } \beta(X) < \text{degree } p(X).$$

Since $p(X)$ and $q(X)$ are in I, also $\beta(X)$ lies in I. The minimality of the degree of $p(X)$ implies $\beta(X) = 0$. Thus $p(X)$ divides $q(X)$ and I is the principal ideal generated by $p(X)$.   □

Proposition 2.5 *Let K be a field and $p(X) \in K[X]$. The following conditions are equivalent:*

1. *the polynomial $p(X)$ is irreducible;*
2. *the ideal $\langle p(X) \rangle$ of $K[X]$ generated by $p(X)$ is maximal;*
3. *the K-algebra $K[X]/\langle p(X) \rangle$ is a field.*

Proof $(1 \Rightarrow 2)$. Consider $\langle p(X) \rangle \subseteq I$ with I an ideal. By Proposition 2.4 we have $I = \langle s(X) \rangle$ for some polynomial $s(X)$. Thus $p(X) \in \langle s(X) \rangle$ and there exists a polynomial $r(X)$ such that $p(X) = r(X) \cdot s(X)$. If $s(X)$ is a non-zero constant, $\langle s(X) \rangle = K[X]$. Otherwise, $r(X)$ is a non-zero constant because $p(X)$ is irreducible; in this case, $\langle p(X) \rangle = \langle s(X) \rangle$.

$(2 \Rightarrow 3)$. Consider the quotient $q \colon K[X] \longrightarrow K[X]/\langle p(X) \rangle$. Each ideal $I \subseteq K[X]/\langle p(X) \rangle$ induces an ideal $q^{-1}(I) \supseteq \langle p(X) \rangle$. Since $\langle p(X) \rangle$ is maximal, $q^{-1}(I) = \langle p(X) \rangle$ or $q^{-1}(I) = K[X]$. Thus $I = qq^{-1}(I) = (0)$ or $I = qq^{-1}(I) = K[X]/\langle p(X) \rangle$. This proves that the ideals of $K[X]/\langle p(X) \rangle$ are trivial and we conclude by Proposition 2.2.

$(3 \Rightarrow 1)$. Let $p(X) = s(X) \cdot r(X)$ be a factorization of $p(X)$. We have $\langle p(X) \rangle \subseteq \langle s(X) \rangle$, thus $\langle s(X) \rangle/\langle p(X) \rangle$ is an ideal of the field $K[X]/\langle p(X) \rangle$. Let us again apply Proposition 2.2. If this ideal is (0), $\langle s(X) \rangle = \langle p(X) \rangle$ and $p(X)$ divides $s(X)$; thus $r(X)$ is a constant. If this ideal is $K[X]/\langle p(X) \rangle$, the equivalence class modulo $p(X)$ of the constant polynomial 1 is in the ideal generated by $s(X)$. This means the existence of a polynomial $u(X)$ such that 1 equals $u(X) \cdot s(X)$ modulo $p(X)$, that is, $1 - u(X) \cdot s(X)$ has the form $v(X) \cdot p(X)$. So there exist polynomials $u(X)$ and $v(X)$ such that

$$1 = u(X) \cdot s(X) + v(X) \cdot p(X) = s(X) \cdot \big(u(X) + v(X) \cdot r(X)\big).$$

This implies that $s(X)$ is a constant. $\square$

Proposition 2.6 *Let K be a field and $n \in \mathbb{N}$ a natural number. Each ideal I of the K-algebra K^n has the form*

$$I = \big\{ (k_i)_{1 \leq i \leq n} \,\big|\, \forall i \in J \ \ k_i = 0 \big\}$$

where $J \subseteq \{1, \ldots, n\}$ is an arbitrary subset of indices.

Proof It is obvious that the subsets I as in the statement are ideals of K^n. Conversely if $I \subseteq K^n$ is an ideal, let us put

$$J = \big\{ j \,\big|\, 1 \leq j \leq n, \ \ \forall (a_i)_{1 \leq i \leq n} \in I \ \ a_j = 0 \big\}.$$

Let us use the notation

$$I_J = \big\{ (a_i)_{1 \leq i \leq n} \,\big|\, \forall j \in J \ \ a_j = 0 \big\}.$$

We have $I \subseteq I_J$ by definition of J, thus it suffices to prove the other inclusion.

For each index $j \notin J$, there exists an element $k^j = (k_i^j)_{1 \leq i \leq n} \in I$ with $k_j^j \neq 0$. Let us write $e_i \in K^n$ to indicate the element whose component of index i is 1, while the other components are 0. Notice that when $j \notin J$, $k^j e_j = k_j^j e_j$. For this reason when $x = (x_i)_{1 \leq i \leq n} \in I_J$,

$$x = \sum_{1 \le i \le n} x_i e_i = \sum_{i \notin J} x_i e_i = \sum_{i \notin J} \frac{x_i e_i}{k_i^i} k_i^i = \sum_{i \notin J} \frac{x_i e_i}{k_i^i} k^i,$$

and this is an element of I because each k^i is in I.   □

Proposition 2.7 *Let K be a field and A a K-algebra. Each homomorphism of K-algebras $f \colon A \longrightarrow K$ is surjective.*

Proof If $k \in K$, we have $k = k \cdot 1 = k \cdot f(1) = f(k \cdot 1)$.   □

Proposition 2.8 (Chinese lemma) *Let K be a field and*

$$(f_i \colon A \longrightarrow K)_{1 \le i \le n}$$

a finite family of distinct homomorphisms of K-algebras. The factorization

$$f \colon A \longrightarrow K^n, \quad a \mapsto \big(f_i(a)\big)_{1 \le i \le n}$$

is surjective.

Proof Each homomorphism $f_i \colon A \longrightarrow K$ is surjective (see Proposition 2.7), thus each f_i is a quotient morphism $f_i \colon A \twoheadrightarrow A/\mathrm{Ker}\, f_i \cong K$. In particular, two distinct homomorphisms f_i, f_j have distinct kernels. By Corollary 2.3, these kernels are maximal ideals of A. Thus $\mathrm{Ker}\, f_i + \mathrm{Ker}\, f_j = A$, for $i \ne j$.

For each pair $i \ne j$ of distinct indices, let us choose $\alpha_{ij} \in \mathrm{Ker}\, f_i$ and $\beta_{ij} \in \mathrm{Ker}\, f_j$ such that $\alpha_{ij} + \beta_{ij} = 1$. We obtain

$$f_j(\alpha_{ij}) = f_j(\alpha_{ij}) + f_j(\beta_{ij}) = f_j(\alpha_{ij} + \beta_{ij}) = f_j(1) = 1.$$

Next, let us put

$$\alpha_j = \prod_{i \ne j} \alpha_{ij}$$

and let us observe that

$$\begin{cases} f_j(\alpha_j) = 1, \\ f_k(\alpha_j) = 0 & \text{if } k \ne j. \end{cases}$$

Let us fix now $(k_i)_{1 \le i \le n} \in K^n$ and, for each index i, let us choose $a_i \in A$ such that $f_i(a_i) = k_i$ (see Proposition 2.7). The element

$$a = \sum_{1 \le k \le n} \alpha_k a_k$$

is such that

$$f_j(a) = \sum_{1 \le k \le n} f_j(\alpha_k) f_j(a_k) = f_j(a_j) = k_j.$$

Thus

$$f(a) = \big(f_j(a)\big)_{1 \le j \le n} = (k_j)_{1 \le j \le n}.$$ □

Theorem 2.9 *Let K be a field and A a K-algebra. The K-algebra homomorphisms $A \longrightarrow K$ are linearly independent in the K-vector space of K-linear mappings from A to K.*

Proof Let $(f_i : A \longrightarrow K)_{1 \leq i \leq n}$ be a finite family of distinct homomorphisms of K-algebras such that $\sum_{1 \leq i \leq n} k_i f_i = 0$, with $k_i \in K$. The Chinese lemma (see Proposition 2.8) implies that the function

$$f : A \longrightarrow K^n, \quad a \mapsto \left(f_i(a_i) \right)_{1 \leq i \leq n}$$

is surjective. If at least one k_i is not zero, $\sum_{1 \leq i \leq n} k_i X_i = 0$ is the equation of a proper K-vector subspace of K^n. Since $\sum_{1 \leq i \leq n} k_i f_i = 0$, this proper subspace contains the image of the function f. This contradicts the surjectivity of f. Thus all elements k_i are zero. $\qquad\qquad\square$

2.2 Extension of Scalars

Let us now recall some classical results concerning the theory of algebras, in the special case of a field extension $K \subseteq L$. The basic result is:

Proposition 2.10 *Let $K \subseteq L$ be a field extension.*

1. *Each L-algebra B is a K-algebra, by restriction of the scalar multiplication.*
2. *For each K-algebra A, $L \otimes_K A$ is an L-algebra for the multiplication*

$$(l \otimes a)(l' \otimes a') = (ll') \otimes (aa')$$

and the scalar multiplication

$$l(l' \otimes a) = (ll') \otimes a,$$

where $l, l' \in L$ and $a, a' \in A$.
3. *These constructions extend as functors*

$$L\text{-Alg} \longrightarrow K\text{-Alg}, \ B \mapsto B, \quad K\text{-Alg} \longrightarrow L\text{-Alg}, \ A \mapsto L \otimes_K A,$$

with the second functor left adjoint to the first one.[1]

Proof Only the adjunction part is not totally trivial. It means the existence of natural isomorphisms

$$\mathsf{Hom}_L(L \otimes_K A, B) \cong \mathsf{Hom}_K(A, B).$$

An L-linear morphism $f : L \otimes_K A \longrightarrow B$ induces a K-linear morphism

$$f' : A \longrightarrow B, \quad a \mapsto f(1 \otimes a)$$

---

[1] Section 5.1 presents a short introduction to the theory of adjoint functors.

while a K-linear morphism $g: A \longrightarrow B$ induces an L-linear morphism

$$g': L \otimes_K A \longrightarrow B, \quad l \otimes a \mapsto lg(a).$$

Observe that $l(f(1 \otimes a)) = f(l \otimes a)$ by the L-linearity of f; this immediately implies the result. ☐

Proposition 2.11 *Let $K \subseteq L$ be a field extension and $p(X) \in K[X]$ an arbitrary polynomial. There exists an isomorphism*

$$L \otimes_K \frac{K[X]}{\langle p(X) \rangle} \cong \frac{L[X]}{\langle p(X) \rangle}$$

where, on the right-hand side, $p(X)$ is considered as a polynomial with coefficients in L.

Proof It is routine to observe that the mappings

$$L \otimes_K \frac{K[X]}{\langle p(X) \rangle} \longrightarrow \frac{L[X]}{\langle p(X) \rangle}, \quad \sum_{i=1}^{n} l_i \otimes [q_i(X)] \mapsto \left[\sum_{i=1}^{n} l_i q_i(X) \right],$$

$$\frac{L[X]}{\langle p(X) \rangle} \longrightarrow L \otimes_K \frac{K[X]}{\langle p(X) \rangle}, \quad \left[\sum_{i=1}^{n} l_i X^i \right] \mapsto \sum_{i=1}^{n} l_i \otimes [X^i],$$

describe the announced isomorphism. ☐

Proposition 2.12 *Let $K \subseteq L$ be a field extension and $p(X) \in K[X]$ an arbitrary polynomial. There exists a bijection between*

1. *the roots of $p(X)$ in L,*
2. *the homomorphisms of K-algebras $\frac{K[X]}{\langle p(X) \rangle} \longrightarrow L$.*

Proof Each element $l \in L$ such that $p(l) = 0$ yields a morphism of K-algebras

$$\mathrm{ev}_l: \frac{K[X]}{\langle p(X) \rangle} \longrightarrow L, \quad [q(X)] \mapsto q(l).$$

Conversely, a morphism $f: \frac{K[X]}{\langle p(X) \rangle} \longrightarrow L$ of K-algebras yields the element $l = f([X])$, where $[X]$ indicates the equivalence class of the polynomial $X \in K[X]$. This element l is a root of $p(X)$ because f, a homomorphism of K-algebras, fixes the elements of K, and thus the coefficients of $p(X)$:

$$p(l) = p\big(f([X])\big) = f\big(p([X])\big) = f\big([p(X)]\big) = f(0) = 0.$$

Moreover, these two constructions are inverse bijections. Indeed, if l is a root of $p(X)$, we get immediately $\mathrm{ev}_l([X]) = l$. Next, beginning with a homomorphism f, for each polynomial $q(X) \in K[X]$,

$$\mathsf{ev}_{f([X])}\Big(\big[q(X)\big]\Big) = q\Big(f([X])\Big) = f\Big(q([X])\Big) = f\Big(\big[q(X)\big]\Big),$$

again because f fixes K, thus the coefficients of $q(X)$.   □

2.3 Algebraic ... Algebras

This section generalizes, to the case of algebras, various results encountered before in the case of field extensions.

Let K be a field and A a K-algebra. For each polynomial

$$p(X) = k_n X^n + k_{n-1} X^{n-1} + \cdots + k_1 X + k_0 \in K[X]$$

and each element $a \in A$, there exists an element

$$p(a) = k_n \cdot a^n + k_{n-1} \cdot a^{n-1} + \cdots + k_1 \cdot a + k_0 \cdot 1 \in A$$

where $1 \in A$ is the unit of the multiplication. When $p(a) = 0$, we shall keep saying that *a is a root of $p(X)$ in A*.

Definition 2.13 Let K be a field. A K-algebra A is *algebraic* over K when each element of A is root of a polynomial $p(X) \in K[X]$.

When no confusion can occur, we shall just write "algebraic" instead of "algebraic over K".

Proposition 2.14 *Let K be a field. Each K-subalgebra $B \subseteq A$ of an algebraic K-algebra A is still algebraic.*

Proof Each element $b \in B \subseteq A$ is root of a polynomial $p(X) \in K[X]$.   □

Proposition 2.15 *Let K be a field. Each finite-dimensional K-algebra is algebraic.*

Proof Same argument as in Proposition 1.5.   □

Proposition 2.16 *Let K be a field and A an algebraic K-algebra. For each element $a \in A$, there exists a unique polynomial $p(X) \in K[X]$ such that*

1. $p(a) = 0$;
2. *the coefficient of the term of maximal degree of $p(X)$ is 1;*
3. *$p(X)$ divides in $K[X]$ each polynomial $q(X)$ such that $q(a) = 0$.*

As a consequence

4. *the degree of $p(X)$ is minimal among the degrees of all polynomials $0 \neq q(X) \in K[X]$ such that $q(a) = 0$.*

This polynomial $p(X)$ is called the minimal polynomial *of a.*

Proof Same proof as for Proposition 1.6.   □

Let us nevertheless observe that in Proposition 2.16, in contrast to Proposition 1.6, we cannot conclude that $p(X)$ is irreducible. Indeed $p(X) = \alpha(X) \cdot \beta(X)$ with $p(a) = 0$ no longer implies that $\alpha(a) = 0$ or $\beta(a) = 0$, because a K-algebra can very well have zero divisors. But of course, the same argument as in Proposition 1.6, still shows that:

Corollary 2.17 *Let K be a field and A an algebraic K-algebra. When the K-algebra A does not have zero divisors, the minimal polynomial of each element $a \in A$ is irreducible.*   □

Example 2.18 (A reducible minimal polynomial) Consider a field K and the K-algebra K^2, of dimension 2 over K, and thus algebraic (see Proposition 2.15). Given $k \in K$, (k, k) is the only root in K^2 of the polynomial $X - k$. In particular since the element $(1, 0) \in K^2$ does not have the form (k, k), its minimal polynomial cannot have degree 1. But $(1, 0)^2 = (1, 0)$, where $(1, 0)$ is a root in K^2 of the polynomial $X^2 - X$. Thus $X^2 - X$ is a polynomial of minimal degree admitting $(1, 0)$ as a root in K^2. The proof of Proposition 2.16 (or Proposition 1.6) implies that this is the minimal polynomial of $(1, 0)$. And of course it is not irreducible, since $X^2 - X = X(X - 1)$. □

Proposition 2.19 *Let K be a field, A an algebraic K-algebra and $0 \neq a \in A$ an element with minimal polynomial $p(X)$ of degree n. The K-subalgebra $K(a) \subseteq A$ generated by a is isomorphic to*

$$K(a) \cong \frac{K[X]}{\langle p(X) \rangle} \cong \left\{ k_0 + k_1 X + \cdots + k_{n-1} X^{n-1} \big| k_i \in K \right\}$$

where, in the last set, the operations are defined modulo $p(X)$.

Proof Trivially, we have

$$K(a) \cong \left\{ q(a) \big| q(X) \in K[X] \right\}.$$

The mapping

$$\left\{ k_0 + k_1 X + \cdots + k_{n-1} X^{n-1} \big| k_i \in K \right\} \longrightarrow K(a), \quad r(X) \mapsto r(a)$$

is surjective. Indeed, if $b = q(a) \in K(a)$, we can divide $q(X)$ by $p(X)$ and get $q(X) = p(X)s(X) + r(X)$, where the degree of $r(X)$ is at most $n - 1$. But $b = q(a) = r(a)$ because $p(a) = 0$.

This mapping is injective as well. Indeed, if the degrees of $r_1(X)$ and $r_2(X)$ are at most $n - 1$, then $r_1(a) = r_2(a)$ implies $(r_1 - r_2)(a) = 0$ with $(r_1 - r_2)(X)$ still of degree at most $n - 1$. The minimality of the degree of $p(X)$ implies $(r_1 - r_2)(X) = 0$, thus $r_1(X) = r_2(X)$.   □

Corollary 2.20 *Let K be a field. Each algebraic K-algebra A, without zero divisors, is a field.*

Proof By Corollary 2.17, with the same arguments as in Propositions 2.19 and 2.5.□

Corollary 2.21 *Let $K \subseteq L$ be an algebraic extension of fields. Each intermediate K-algebra $K \subseteq A \subseteq L$ is still a field.*

Proof The field L does not have zero divisors, thus $A \subseteq L$ does not have zero divisors. We conclude by Proposition 1.4 and Corollary 2.20.   □

Proposition 2.22 *Let K be a field. Each algebraic K-algebra, which is finitely generated as a K-algebra, has finite dimension as a K-vector space.*

Proof Let us consider an algebraic K-algebra A generated by the elements $a_1, \ldots, a_n$. Clearly,
$$A = \left\{ p(a_1, \ldots, a_n) \middle| p(X_1, \ldots, X_n) \in K[X_1, \ldots, X_n] \right\}.$$
But if a_i admits the minimal polynomial $p_i(X)$ of degree n_i,
$$p_i(a_i) = a_i^{n_i} + k_{n_i-1} a_i^{n_i-1} + \cdots + k_1 a_i + k_0 1 = 0,$$
we can substitute each power $a_i^{n_i}$ by an expression of degree $n_i - 1$ in a_i. Since the number of generators is finite, we can also choose a unique integer $n \in \mathbb{N}$ greater than each n_i. In the description of A, it suffices now to work with polynomials $p(X)$ of degree at most n; these constitute a finite-dimensional K-vector space. Thus A also has finite dimension as a K-vector space.   □

2.4 Finite-Dimensional Split Algebras

This time we generalize to the case of algebras some results concerning Galois extensions of fields. The corresponding algebras are called *split algebras*: this notion is the key to generalizing the Galois theorem to the case of rings. For example, observe that Propositions 2.25 and 2.27 translate the notion of Galois extension in terms of a tensor product, removing any reference to the theory of polynomials over a field.

Definition 2.23 Let $K \subseteq L$ be a field extension. A K-algebra A is *split* by the extension $K \subseteq L$ when each element $a \in A$ is root of a polynomial $q(X) \in K[X]$ which admits in $L[X]$ a decomposition into distinct linear factors.

Again, "distinct linear factors" means distinct when written in the canonical form $X - l$. When no confusion can occur, we shall simply say "split by L" instead of "split by the extension $K \subseteq L$".

This definition admits an equivalent formulation:

Proposition 2.24 *Let $K \subseteq L$ be a field extension. A K-algebra A is split by L when*

1. *A is an algebraic K-algebra;*
2. *the minimal polynomial $p(X) \in K[X]$ of each element $a \in A$ admits in $L[X]$ a decomposition into distinct linear factors.*

Proof Clearly, Proposition 2.14 implies Definition 2.23. Conversely, with the notation of these statements, $p(X)$ is a factor of $q(X)$ (see Proposition 2.16), thus admits in $L[X]$ a decomposition into distinct linear factors. □

It suffices now to compare Definitions 1.12 and 2.23 to conclude that:

Proposition 2.25 *Let $K \subseteq L$ be a field extension. The following conditions are equivalent:*

1. *$K \subseteq L$ is a Galois extension;*
2. *L is a K-algebra split by L.* □

Proposition 2.26 *Let $K \subseteq L$ be a field extension and A a K-algebra split by L. Each K-subalgebra $B \subseteq A$ is still split by L.*

Proof Each element $b \in B \subseteq A$ is root of a polynomial $p(X) \subseteq K[X]$ admitting in $L[X]$ a decomposition into distinct linear factors. □

Theorem 2.27 *Let $K \subseteq L$ be a field extension of finite dimension m and A a K-algebra of finite dimension n. The following conditions are equivalent:*

1. *the K-algebra A is split by L;*
2. *the mapping*

$$\mathrm{Gel} \colon L \otimes_K A \longrightarrow L^{\mathrm{Hom}_L(L \otimes_K A, L)},$$
$$l \otimes a \mapsto \big(f(l \otimes a)\big)_{f \in \mathrm{Hom}_L(L \otimes_K A, L)}$$

is an isomorphism of L-algebras, called the Gelfand transformation *of A;*
3. *the mapping*

$$\mathrm{Gel} \colon L \otimes_K A \longrightarrow L^{\mathrm{Hom}_K(A, L)},$$
$$l \otimes a \mapsto \big(lg(a)\big)_{g \in \mathrm{Hom}_K(A, L)}$$

is an isomorphism of L-algebras;
4. *$\#\mathrm{Hom}_L(L \otimes_K A, L) = n$;*
5. *$\#\mathrm{Hom}_K(A, L) = n$;*
6. *$L \otimes_K A$ is isomorphic to L^n as an L-algebra;*
7. *$\forall x \in L \otimes_K A, \ x \neq 0, \ \exists f \in \mathrm{Hom}(L \otimes_K A, L)$ such that $f(x) \neq 0$.*

In these formulæ, # is the cardinality symbol.

Proof Proposition 2.15 implies that A is an algebraic K-algebra. The K-vector space $L \otimes_K A$ has dimension mn over K. The vector space of K-linear mappings $L \otimes_K A \longrightarrow L$ has the finite dimension $(mn)m$ over K. Thus Theorem 2.9 implies that

$\text{Hom}_L(L \otimes_K A, L)$ is finite. Next Theorem 2.9 and the Chinese Lemma (Proposition 2.8) imply that the Gelfand transformation is surjective.

For reasons of clarity, we split the rest of the proof into three lemmas.

Lemma 2.28 *Conditions 2 to 7 are equivalent.*

Proof The equivalences $(2 \Leftrightarrow 3)$ and $(4 \Leftrightarrow 5)$ follow at once from Proposition 2.10.

$(2 \Rightarrow 4)$ and $(2 \Rightarrow 6)$. The K-vector space $L \otimes_K A$ has dimension mn and the K-vector space $L^{\text{Hom}(L \otimes_K A, L)}$ has dimension $m \cdot \#\text{Hom}(L \otimes_K A, L)$. When the Gelfand transformation is an isomorphism, the equality of these dimensions forces $n = \#\text{Hom}(L \otimes_K A, L)$.

$(4 \Rightarrow 2)$. We know already that the Gelfand transformation is surjective. When its domain and codomain have the same finite dimension mn, the Gelfand transformation is an isomorphism.

$(6 \Rightarrow 4)$. We have now $L \otimes_K A \cong L^n$. The projections $p_i \colon L^n \longrightarrow L$ are n distinct homomorphisms of L-algebras; these homomorphisms are linearly independent over L (see Theorem 2.9). But the L-vector space $\text{Lin}_L(L^n, L)$ of all L-linear mappings has dimension n. Again, Theorem 2.9 implies that these n homomorphisms p_i are all the L-algebra homomorphisms $L^n \longrightarrow L$, thus all the homomorphisms $L \otimes_K A \longrightarrow L$.

$(2 \Leftrightarrow 7)$. Condition 7 expresses precisely the injectivity of the Gelfand transformation which, as we already know, is surjective. $\quad\square$

Lemma 2.29 *The class of those K-algebras satisfying the equivalent conditions 2 to 7 is closed under subobjects, quotients, finite products, tensor products and finite unions, in the category of all K-algebras.*

Proof Condition 7 implies the stability under subobjects.

Let $A \longrightarrow\!\!\!\!\rightarrow Q$ be a quotient of K-algebras, with A of dimension n, satisfying conditions 2 to 7. The tensor product with L preserves quotients, because it admits a right adjoint (see Proposition 2.10). We thus get a quotient of L-algebras

$$L^n \cong L \otimes_K A \longrightarrow\!\!\!\!\rightarrow L \otimes_K Q.$$

The kernel of this quotient is an ideal $J \lhd L^n$ which has the form (see Proposition 2.6)

$$J = \left\{ (l_i)_{1 \le i \le n} \middle| \forall i \in X \ l_i = 0 \right\}, \quad X \subseteq \{1, \dots, n\}.$$

Putting $x = \#X$, we obtain $L \otimes_K Q \cong L^n/J \cong L^{n-x}$. By condition 6, it suffices to prove that Q has dimension $n - x$ over K. Since L has dimension m over K and L^n/J has dimension $n - x$ over L, we know already that L^n/J has dimension $m(n - x)$ over K. But $L \otimes_K Q$ has dimension $m \cdot \dim_K Q$ over K, thus $\dim_K Q = n - x$ because $L \otimes_K Q \cong L^n/J$.

To prove the stability under finite products, let us observe that the functor "tensor product" between categories of vector spaces,

$$L \otimes_K - \colon \text{Vect}_K \longrightarrow \text{Vect}_L,$$

is an additive functor, thus preserves finite products of vector spaces. But the product of two algebras is their product as vector spaces, provided with the componentwise multiplication. Thus tensoring with L preserves finite products of algebras. If A, A' are K-algebras of dimensions n, n' satisfying conditions 2 to 7, $A \times A'$ has dimension $n + n'$ and

$$L \otimes_K (A \times A') \cong (L \otimes_K A) \times (L \otimes_K A') \cong L^n \times L^{n'} \cong L^{n+n'}.$$

Thus $A \times A'$ satisfies condition 6.

In the case of the tensor product, $A \otimes_K A'$ has dimension nn'. Since the tensor product preserves finite products, we get

$$L \otimes_K (A \otimes_K A') \cong (L \otimes_K A) \otimes_K A' \cong L^n \otimes_K A'$$
$$\cong (L \otimes_K A')^n \cong \left(L^{n'} \right)^n \cong L^{nn'}.$$

Again $A \otimes_K A'$ satisfies condition 6.

The union of two subalgebras $A_1 \subseteq A$ and $A_2 \subseteq A$, that is the smallest subalgebra $A' \subseteq A$ containing A_1 and A_2, is simply

$$A_1 \cdot A_2 = \left\{ \sum_{i=1}^{k} a_{i,1} a_{i,2} \middle| a_{i,1} \in A_1, \ a_{i,2} \in A_2 \right\}.$$

This algebra is the quotient

$$A_1 \otimes_K A_2 \longrightarrow\!\!\!\!\!\longrightarrow A_1 \cdot A_2, \quad a_1 \otimes a_2 \mapsto a_1 a_2,$$

so that the result follows from the previous parts of the proof of this lemma. $\square$

Lemma 2.30 *Conditions 1 and 2 are equivalent.*

Proof $(1 \Rightarrow 2)$. Consider $a \in A$ with minimal polynomial $p(X) \in K[X]$ of degree r. In L, $p(X)$ admits r distinct roots (see Proposition 2.24), thus $\#\mathsf{Hom}_K (K(a), L) = r$ (see Proposition 2.19 and Corollary 2.12) and $K(a) \cong \frac{K[X]}{\langle p(X) \rangle}$ satisfies condition 5. But A is finite-dimensional over K; thus A is the union of finitely many sub-K-algebras of the form $K(a_i)$. We thus conclude by applying Lemma 2.29 several times.

$(2 \Rightarrow 1)$. Let $p(X) \in K[X]$ be the minimal polynomial, of degree n, of an element $a \in A$. Proposition 2.19 implies that $\frac{K[X]}{\langle p(X) \rangle} \cong K(a)$; in particular, this vector space has dimension n over K. But $K(a) \subseteq A$ satisfies condition 2 by assumption on A and Lemma 2.29 (stability under subobjects). Lemma 2.28 implies now that $\#\mathsf{Hom}_K \left(\frac{K[X]}{\langle p(X) \rangle}, L \right) = n$. By Proposition 2.12, $p(X)$ has exactly n distinct roots in L. $\square$

2.5 The Theory of G-Sets

Let us first recall some less known aspects of group theory.

Definition 2.31 Let $(G, \circ)$ be a group. A G-set is a pair $(X, \cdot)$ where X is a set provided with a G-action

$$\cdot : G \times X \longrightarrow X, \quad (g, x) \mapsto g \cdot x$$

satisfying the axioms

$$1 \cdot x = x, \quad (g \circ g') \cdot x = g \cdot (g' \cdot x)$$

for all $g, g' \in G$ and $x \in X$.

A morphism $f : (X, \cdot) \longrightarrow (Y, \cdot)$ of G-sets is a mapping $f : X \longrightarrow Y$ such that $f(g \cdot x) = g \cdot f(x)$ for all elements $g \in G$ and $x \in X$.

We write G-Set to indicate the category of G-sets and their morphisms.

We shall generally just write gg' and gx instead of $g \circ g'$ and $g \cdot x$. The group G, with the action $g \cdot g' = g \circ g'$, is an example of a G-set.

If $H \subseteq G$ is a subgroup, the quotient G/H is generally not a group (except when H is normal), but it is always a G-set for the multiplication $g \cdot [g'] = [g \circ g']$. More precisely:

Proposition 2.32 *Let G be a group. There exists a bijection between*

- *the subgroups of G,*
- *the quotients of the G-set G.*

Via this bijection, a subgroup $H \subseteq G$ corresponds to the quotient G/H.

Proof Each subgroup $H \subseteq G$ determines an equivalence relation

$$x \equiv y \quad \text{iff} \quad x^{-1}y \in H;$$

where G/H indicates the quotient. This quotient is a G-set for the action $g[x] = [gx]$, where $g, x \in G$.

First, observe that this definition is independent of the choice of the element $x \in [x]$. Indeed, if $x \equiv y$,

$$(gx)^{-1}(gy) = x^{-1}g^{-1}gy = x^{-1}y \in H$$

thus $[gx] = [gy]$.

The G-set axioms are obviously satisfied and the projection $G \longrightarrow G/H$ is a morphism of G-sets.

Observe next that $1 \equiv x$ precisely when $x \in H$, thus when $H = [1]$. Conversely, let $p : G \longrightarrow Q$ be a quotient of G-sets. The equivalence class of $[1] \subseteq G$ is a subgroup of G. Indeed if $x, y \in [1]$,

$$[x^{-1}] = [x^{-1}1] = x^{-1}[1] = x^{-1}[x] = [x^{-1}x] = [1],$$
$$[xy] = x[y] = x[1] = [x1] = [x] = [1].$$

To conclude, it remains to observe that the quotient Q of G is precisely the quotient corresponding to the subgroup $[1]$. Indeed, if $[x] = [y]$,

$$[x^{-1}y] = x^{-1}[y] = x^{-1}[x] = [x^{-1}x] = [1].$$

Conversely if $x^{-1}y \in [1]$,

$$[y] = [xx^{-1}y] = x[x^{-1}y] = x[1] = [x1] = [x]. \qquad\qquad \square$$

Proposition 2.33 *Let G be a group. Every G-set X is a disjoint union of quotients of the G-set G.*

Proof It is immediate to observe that a disjoint union of G-sets, with the original actions of G on each term of this union, is still a G-set. If $x \in X$, we have a surjection

$$G \longrightarrow\!\!\!\!\!\!\twoheadrightarrow Gx = \{gx | g \in G\}, \quad Gx \subseteq X.$$

The set Gx is a G-subset of X and, by definition, is isomorphic to a quotient of the G-set G. If $y \in X \setminus Gx$, necessarily Gx and Gy are disjoint. Indeed, $gx = g'y$ would imply $y = (g')^{-1}gx \in Gx$. This forces the conclusion. $\qquad\qquad \square$

2.6 The Galois Theorem of Grothendieck

This section proposes a proof of *Grothendieck's version* of the Galois theorem, using only classical techniques of field and algebra theories. It refers to the following notion:

Definition 2.34 A functor $F \colon \mathcal{A} \longrightarrow \mathcal{B}$ is an *equivalence* of categories when there exists a functor $G \colon \mathcal{B} \longrightarrow \mathcal{A}$ and natural isomorphisms $G \circ F \cong \mathrm{id}_{\mathcal{A}}$, $F \circ G \cong \mathrm{id}_{\mathcal{B}}$.

Some more terminology will make the language easier:

Definition 2.35 Let $F \colon \mathcal{A} \longrightarrow \mathcal{B}$ be a functor. For each pair of objects $A, A' \in \mathcal{A}$ consider the mapping

$$\varphi_{A,A'} \colon \mathcal{A}(A, A') \longrightarrow \mathcal{B}\big(F(A), F(A')\big), \quad a \mapsto F(a).$$

- the functor F is *faithful* when all the mappings $\varphi_{A,A'}$ are injective;
- the functor F is *full* when all the mappings $\varphi_{A,A'}$ are surjective;
- the functor F is *essentially surjective on the objects* when each object $B \in \mathcal{B}$ is isomorphic to an object $F(A)$, with $A \in \mathcal{A}$.

Lemma 2.36 *A functor* $F\colon \mathcal{A} \longrightarrow \mathcal{B}$ *is an equivalence of categories if and only if it is faithful, full and essentially surjective on the objects.*

Proof An equivalence $F\colon \mathcal{A} \longrightarrow \mathcal{B}$ is essentially surjective on the objects since every $B \in \mathcal{B}$ is isomorphic to $FG(B)$. Moreover, using the notation of Definition 2.35 and writing further

$$\gamma_{B,B'}\colon \mathcal{B}(B, B') \longrightarrow \mathcal{A}\big(G(B), G(B')\big), \quad b \mapsto G(b)$$

the action of the composite

$$\mathcal{A}(A, A') \xrightarrow{\ \varphi_{A,A'}\ } \mathcal{B}\big(F(A), F(A')\big) \xrightarrow{\ \gamma_{F(A),F(A')}\ } \mathcal{A}\big(GF(A), GF(A')\big)$$

is, by assumption, naturally isomorphic to the action of the identity, thus is bijective. Therefore $\varphi_{A,A'}$ is injective and $\gamma_{F(A),F(A')}$ is surjective. But since each object $B \in \mathcal{B}$ is isomorphic to an object of the form $F(A)$, each $\gamma_{B,B'}$ is surjective as well. Interchanging the roles of F and G, we conclude that each $\gamma_{B,B'}$ is injective and each $\varphi_{A,A'}$ is surjective. So we have bijections in both cases.

Conversely, assume that F is full, faithful and essentially surjective on the objects. Using the axiom of choice, for each object $B \in \mathcal{B}$, choose an object $G(B) \in \mathcal{A}$ and an isomorphism $\alpha_B\colon B \cong FG(B)$. For each arrow $b\colon B \longrightarrow B'$ one gets the following situation

$$
\begin{array}{ccc}
G(B) & \qquad G \xrightarrow[\alpha_B]{\cong} FG(B) \\[4pt]
\Big\downarrow{\scriptstyle G(b)} & \quad b\Big\downarrow \qquad \qquad \Big\downarrow{\scriptstyle \alpha_{B'} \circ b \circ \alpha_B^{-1}} \\[4pt]
G(B') & \qquad B' \xrightarrow[\alpha_{B'}]{\cong} FG(B')
\end{array}
$$

and one can define $G(b)$ to be the morphism corresponding to $\alpha_{B'} \circ b \circ \alpha_B^{-1}$ by the bijection

$$\varphi_{G(B),G(B')}\colon \mathcal{A}\big(G(B), G(B')\big) \longrightarrow \mathcal{B}\big(FG(B), FG(B')\big), \quad a \mapsto F(a).$$

Observe next that given $A \in \mathcal{A}$, we have the bijection

$$\varphi_{A,GF(A)}\colon \mathcal{A}\big(A, GF(A)\big) \longrightarrow \mathcal{B}\big(F(A), FGF(A)\big).$$

The morphism $\alpha_{F(A)}\colon F(A) \longrightarrow FGF(A)$ corresponds via this bijection to a morphism $\beta_A\colon A \longrightarrow GF(A)$. The inverse isomorphism $\alpha_{F(A)}^{-1}$ will then yield the inverse of β_A via the bijection $\varphi_{GF(A),A}$.

Checking the details is straightforward. $\square$

Theorem 2.37 (Galois Theorem) *Let $K \subseteq L$ be a finite-dimensional Galois extension of fields and* $\mathsf{Gal}[L : K]$ *the corresponding Galois group. We consider the functor*

$$\mathsf{Hom}_K(-, L) \colon \mathsf{Split}[L : K]_f \longrightarrow \mathsf{Gal}[L : K]\text{-}\mathsf{Set}_f,$$
$$A \mapsto \mathsf{Hom}_K(A, L)$$

where

1. $\mathsf{Split}[L : K]_f$ *is the category of finite-dimensional K-algebras split by L;*
2. $\mathsf{Gal}[L : K]\text{-}\mathsf{Set}_f$ *is the category of finite* $\mathsf{Gal}[L : K]$*-sets;*
3. $\mathsf{Hom}_K(A, L)$ *is the set of K-algebra homomorphisms;*
4. *the action of* $\mathsf{Gal}[L : K]$ *over* $\mathsf{Hom}_K(A, L)$

$$\mathsf{Gal}[L : K] \times \mathsf{Hom}_K(A, L) \longrightarrow \mathsf{Hom}_K(A, L),$$
$$(g, f) \mapsto g \circ f$$

is simply composition.

This functor is a contravariant equivalence of categories (see Definition 2.34).

It is obvious that $\mathsf{Hom}_K(-, L)$, as defined in the statement, is a contravariant functor. To make the proof easier to follow, we split it into five lemmas.

Lemma 2.38 *For each algebra $A \in \mathsf{Split}[L : K]_f$,*

$$\mathsf{Gal}[L : K] \times (L \otimes_K A) \longrightarrow L \otimes_K A,$$
$$\left(g, \sum_i l_i \otimes a_i \right) \mapsto \sum_i g(l_i) \otimes a_i$$

defines the structure of a $\mathsf{Gal}[L : K]$*-set on $L \otimes_K A$. Via the Gelfand isomorphism (see Theorem 2.27), this action becomes:*

$$\mathsf{Gal}[L : K] \times L^{\mathsf{Hom}_K(A,L)} \longrightarrow L^{\mathsf{Hom}_K(A,L)},$$
$$(g, \varphi) \mapsto \left[f \mapsto g\big(\varphi(g^{-1} \circ f)\big) \right],$$

where $\varphi \colon \mathsf{Hom}_K(A, L) \longrightarrow L$ and $f \in \mathsf{Hom}_K(A, L)$.

Proof It is obvious that the statement defines a $\mathsf{Gal}[L : K]$-set structure on $L \otimes_K A$. Let us now fix $g \in \mathsf{Gal}[L : K]$ and consider the morphism

$$\gamma \colon L^{\mathsf{Hom}_K(A,L)} \longrightarrow L^{\mathsf{Hom}_K(A,L)}, \quad (\gamma(\varphi))(f) = g\big(\varphi(g^{-1} \circ f)\big).$$

We get

$$\Big((\gamma \circ \mathsf{Gel})(l \otimes a)\Big)(f) = \Big(\gamma\big(\mathsf{Gel}(l \otimes a)\big)\Big)(f)$$
$$= g\Big(\mathsf{Gel}(l \otimes a)(g^{-1} \circ f)\Big)$$
$$= g\Big(l\big((g^{-1} \circ f)(a)\big)\Big)$$
$$= g\Big(lg^{-1}\big(f(a)\big)\Big)$$
$$= g(l)gg^{-1}\big(f(a)\big)$$
$$= g(l)f(a)$$
$$= \mathsf{Gel}\Big((g \otimes \mathsf{id})(l \otimes a)\Big)(f)$$
$$= \Big((\mathsf{Gel} \circ (g \otimes \mathsf{id}))(l \otimes a)\Big)(f).$$

Thus the following diagram is commutative

$$
\begin{array}{ccc}
L \otimes_K A & \xrightarrow[\cong]{\mathsf{Gel}} & L^{\mathsf{Hom}_K(A,L)} \\
{\scriptstyle g \otimes \mathsf{id}} \downarrow & & \downarrow {\scriptstyle \gamma} \\
L \otimes_K A & \xrightarrow[\mathsf{Gel}]{\cong} & L^{\mathsf{Hom}_K(A,L)}
\end{array}
$$

This commutativity expresses precisely the equivalence between the two formulæ in the statement. □

Lemma 2.39 *For each algebra* $A \in \mathsf{Split}[L : K]_f$,

$$A \cong \big\{x \in L \otimes_K A \,\big|\, \forall g \in \mathsf{Gal}[L : K]\ (g \otimes \mathsf{id})(x) = x\big\}.$$

Proof We can identify A with a subset of $L \otimes_K A$ via the inclusion

$$A \cong K \otimes_K A \rightarrowtail L \otimes_K A, \quad a \mapsto 1 \otimes a.$$

Indeed, as a K-vector subspace of L, K admits a supplementary K-vector subspace K', allowing us to write $L \cong K \oplus K'$ as a K-vector space. The inclusion $i \colon K \longrightarrow L$ thus admits the projection $p \colon K \oplus K' \longrightarrow K$ as a K-linear retraction. The morphism $i \otimes_K \mathsf{id}_A$ above then admits $p \otimes_R \mathsf{id}_A$ as a K-linear retraction, and is therefore injective.

Moreover for each $g \in \mathsf{Gal}[L : K]$, we have $(g \otimes \mathsf{id})(1 \otimes a) = 1 \otimes a$; this proves already that

$$A \subseteq \big\{x \in L \otimes_K A \,\big|\, \forall g \in \mathsf{Gal}[L : K]\ (g \otimes \mathsf{id})(x) = x\big\}.$$

To prove the equality, let us observe that the finite-dimensional K-algebra A, seen as a K-vector space, is isomorphic to a power K^n, $n \in \mathbb{N}$. Consider the commutative diagram

$$L \otimes_K K^n \xrightarrow{\quad g \otimes \text{id} \quad} L \otimes_K K^n$$

$$\cong \Big\downarrow \qquad\qquad \Big\downarrow \cong$$

$$L^n \xrightarrow{\quad g^n \quad} L^n$$

for each $g \in \text{Gal}[L : K]$. This reduces the problem to the consideration of those elements of L^n fixed by each g^n. Using the classical Galois theorem (see Theorem 1.24), we obtain

$$\left\{ x \in L \otimes_K K^n \,\middle|\, \forall g \in \text{Gal}[L : K] \ (g \otimes \text{id})(x) = x \right\} \cong \Big(\text{Fix}(\text{Gal}[L : K]) \Big)^n \cong K^n.$$

The form of the isomorphism

$$L^n \xrightarrow{\quad \cong \quad} L \otimes_K K^n, \quad (l_i)_{1 \leq i \leq n} \mapsto \sum_{i=1}^n l_i \otimes e_i,$$

where the e_i are the vectors of the canonical basis of K^n, implies

$$\left\{ x \in L \otimes_K K^n \,\middle|\, \forall g \in \text{Gal}[L : K] \ (g \otimes \text{id})(x) = x \right\}$$

$$\cong \left\{ \sum_{i=1}^n k_i \otimes e_i \,\middle|\, k_i \in K \right\} \cong \left\{ 1 \otimes \left(\sum_{i=1}^n k_i e_i \right) \,\middle|\, k_i \in K \right\} \cong A.$$

This concludes the proof of the lemma. $\qquad\qquad\qquad\qquad\qquad\qquad\qquad\qquad\qquad\quad$ □

Lemma 2.40 *The functor* $\text{Hom}_K(-, L)$ *is full (see Definition 2.35).*

Proof Let us fix two K-algebras A and B in $\text{Split}[L : K]_f$ and a morphism of $\text{Gal}[L : K]$-sets

$$\varphi : \text{Hom}_K(B, L) \longrightarrow \text{Hom}_K(A, L).$$

This induces a mapping

$$L^\varphi : L^{\text{Hom}_K(A,L)} \longrightarrow L^{\text{Hom}_K(B,L)},$$

$$(l_f)_{f \in \text{Hom}_K(A,L)} \mapsto (l_{\varphi(h)})_{h \in \text{Hom}_K(B,L)}.$$

Lemma 2.38 describes the $\text{Gal}[L : K]$-set structure of these powers of L. Let us observe that L^φ is a morphism of $\text{Gal}[L : K]$-sets. Indeed, take $g \in \text{Gal}[L : K]$; to avoid a possible ambiguity between the image by the morphism g and an action of the element g, we shall indicate the action of g with the symbol $*$. We have

$$L^{\varphi}\left(g * (l_f)_{f \in \mathsf{Hom}_K(A,L)}\right) = L^{\varphi}\left(g(l_{g^{-1} \circ f})\right)_{f \in \mathsf{Hom}_K(A,L)}$$
$$= \left(g(l_{g^{-1} \circ \varphi(h)})\right)_{h \in \mathsf{Hom}_K(B,L)}$$
$$= g * \left(l_{\varphi(h)}\right)_{h \in \mathsf{Hom}_K(B,L)}$$
$$= g * L^{\varphi}\left((l_f)_{f \in \mathsf{Hom}_K(A,L)}\right).$$

But L^{φ} is a morphism of $\mathsf{Gal}[L : K]$-sets, thus it admits a restriction through the $\mathsf{Gal}[L : K]$-set of those elements fixed by the action of all elements g. By Lemma 2.39 and using the Gelfand isomorphisms (see Theorem 2.27), we obtain the situation

$$
\begin{array}{ccc}
A & \xrightarrow{\;\cong\;} \mathsf{FIX}(L \otimes_K A) \xrightarrow{\;\cong\;} \mathsf{FIX}\left(L^{\mathsf{Hom}_K(A,L)}\right) \\
& \Big\downarrow{\scriptstyle L^{\varphi}} \\
B & \xleftarrow{\;\cong\;} \mathsf{FIX}(L \otimes_K B) \xleftarrow{\;\cong\;} \mathsf{FIX}\left(L^{\mathsf{Hom}_K(B,L)}\right)
\end{array}
$$

where FIX indicates the set of those elements fixed by the action of each element $g \in \mathsf{Gal}[L : K]$. Let us write $\psi \colon A \longrightarrow B$ for this composite.

Let us now prove that $\varphi = \mathsf{Hom}_K(\psi, L)$. If $h \in \mathsf{Hom}_K(B, L)$, let us consider the diagram:

$$
\begin{array}{ccc}
L \otimes_K B & \xrightarrow{\;\mathsf{Gel}_B\;}_{\cong} & L^{\mathsf{Hom}_K(B,L)} \\
{\scriptstyle i_B}\Big\uparrow & & \Big\downarrow{\scriptstyle p_h} \\
B & \xrightarrow{\;\;h\;\;} & L
\end{array}
$$

where $i_B(b) = 1 \otimes b$ e p_h is the projection of index h. This diagram is commutative because

$$(p_h \circ \mathsf{Gel}_B \circ i_B)(b) = (p_h \circ \mathsf{Gel}_B)(1 \otimes b) = p_h\left(h'(b)_{h' \in \mathsf{Hom}_K(B,L)}\right) = h(b).$$

Let us further write $\overline{\varphi} \colon L \otimes_K B \longrightarrow L \otimes_K B$ to indicate the morphism corresponding to L^{φ} by the Gelfand isomorphism. The definitions of ψ and $\overline{\varphi}$ imply the commutativity of the diagram

$$
\begin{array}{ccccc}
A & \xrightarrow{\;i_A\;} & L \otimes_K A & \xrightarrow{\;\mathsf{Gel}_A\;}_{\cong} & L^{\mathsf{Hom}_K(A,L)} \\
{\scriptstyle \psi}\Big\downarrow & & {\scriptstyle \overline{\varphi}}\Big\downarrow & & \Big\downarrow{\scriptstyle L^{\varphi}} \\
B & \xrightarrow{\;i_B\;} & L \otimes_K B & \xrightarrow[\;\mathsf{Gel}_B\;]{\cong} & L^{\mathsf{Hom}_K(B,L)}
\end{array}
$$

This implies further

$$
\begin{aligned}
\mathrm{Hom}_K(\psi, L)(h) = h \circ \psi \\
= p_h \circ \mathrm{Gel}_B \circ i_B \circ \psi \\
= p_h \circ \mathrm{Gel}_B \circ \overline{\varphi} \circ i_A \\
= p_h \circ L^\varphi \circ \mathrm{Gel}_A \circ i_A \\
= p_{\varphi(h)} \circ \mathrm{Gel}_A \circ i_A \\
= \varphi(h),
\end{aligned}
$$

which concludes the proof of the lemma.   □

Lemma 2.41 *The functor* $\mathrm{Hom}_K(-, L)$ *is faithful (see Definition 2.35).*

Proof With the notation of Lemma 2.40, consider another morphism $\psi' \colon A \longrightarrow B$ such that $\mathrm{Hom}_K(\psi', L) = \varphi$. For each $h \in \mathrm{Hom}_K(B, L)$ we get

$$
\begin{aligned}
p_h \circ \mathrm{Gel}_B \circ i_B \circ \psi' = h \circ \psi' \\
= \varphi(h) \\
= p_{\varphi(h)} \circ \mathrm{Gel}_A \circ i_A \\
= p_h \circ L^\varphi \circ \mathrm{Gel}_A \circ i_A \\
= p_h \circ \mathrm{Gel}_B \circ i_B \circ \psi.
\end{aligned}
$$

This equality is valid for each projection p_h, thus

$$
\mathrm{Gel}_B \circ i_B \circ \psi' = \mathrm{Gel}_B \circ i_B \circ \psi.
$$

Since Gel_B and i_B are injective, $\psi' = \psi$.   □

Lemma 2.42 *The functor* $\mathrm{Hom}_K(-, L)$ *is essentially surjective on the objects (see Definition 2.35).*

Proof First, consider a subgroup $H \subseteq \mathrm{Gal}[L : K]$ and the corresponding quotient $\mathrm{Gal}[L : K]/H$ in the category of $\mathrm{Gal}[L : K]$-sets (see Proposition 2.32). All these objects are finite, because $\mathrm{Gal}[L : K]$ is finite (see Theorem 1.24). Let us prove that

$$
\frac{\mathrm{Gal}[L : K]}{H} \cong \mathrm{Hom}_K\big(\mathrm{Fix}(H), L\big).
$$

Consider the inclusion $\mathrm{Fix}(H) \subseteq L$ and the corresponding morphism of $\mathrm{Gal}[L : K]$-sets

$$
\rho \colon \mathrm{Gal}[L : K] \cong \mathrm{Hom}_K(L, L) \longrightarrow \mathrm{Hom}_K\big(\mathrm{Fix}(H), L\big)
$$

which maps $f \colon L \longrightarrow L$ to its restriction $f \colon \mathrm{Fix}(H) \longrightarrow L$. Considering the diagram

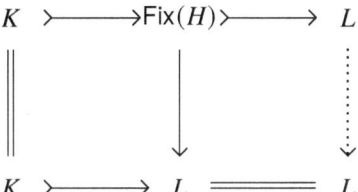

and Proposition 1.16, we infer that each morphism of K-algebras $\mathsf{Fix}(H)\longrightarrow L$ is the restriction of a morphism $L\longrightarrow L$. Thus ρ is a quotient morphism.

To prove that $\mathsf{Hom}_K(\mathsf{Fix}(H), L)$ is isomorphic to the quotient $\mathsf{Gal}[L : K]/H$, we must still prove that two morphisms of K-algebras $f, g\colon L \rightrightarrows L$ have the same restriction on $\mathsf{Fix}(H)$ precisely when $f^{-1}\circ g\in H$. Indeed, f and g have the same restriction on $\mathsf{Fix}(H)$ precisely when $f^{-1}\circ g$ fixes the elements of $\mathsf{Fix}(H)$, that is, using the classical Galois theorem (see Theorem 1.24), when

$$f^{-1}\circ g\in \mathsf{Gal}\big[L : \mathsf{Fix}(H)\big] = H.$$

Thus each quotient of the $\mathsf{Gal}[L : K]$-set $\mathsf{Gal}[L : K]$ is isomorphic to an object of the form $\mathsf{Hom}_K(A, L)$ for some object $A\in \mathsf{Split}[L : K]_f$. But each finite $\mathsf{Gal}[L : K]$-set is a finite disjoint union of quotients of $\mathsf{Gal}[L : K]$ (see Proposition 2.33). Since the category $\mathsf{Split}[L : K]_f$ has finite products (see Lemma 2.29), it remains now to prove that the functor $\mathsf{Hom}_K(-, L)$ transforms finite products into finite disjoint unions.

To prove this, consider two K-algebras $A, B\in \mathsf{Split}[L : K]_f$ of respective dimensions n and m. The composites with the projections

$$A\twoheadleftarrow\!\!\!\!\!\!-\!\!\!-\; A\times B \;-\!\!\!-\!\!\!\!\!\!\twoheadrightarrow B$$

yield morphisms

$$\mathsf{Hom}_K(A, L)\rightarrowtail\!\!\!-\!\!\!\!\longrightarrow \mathsf{Hom}_K(A\times B, L)\longleftarrow\!\!\!-\!\!\!\!\leftarrowtail \mathsf{Hom}_K(B, L)$$

which are injective, because the projections are surjective. These subobjects are disjoint. Indeed each $f\in \mathsf{Hom}_K(A, L)$ is such that $(f\circ p_A)(1, 0) = f(1) = 1$ and each $g\in \mathsf{Hom}_K(B, L)$ is such that $(g\circ p_B)(1, 0) = g(0) = 0$. Theorem 2.27 and various arguments already used in this proof imply

$$
\begin{aligned}
\#\mathsf{Hom}_K(A\times B, L) &= \#\mathsf{Hom}_L\big(L\otimes_K (A\times B), L\big)\\
&= \#\mathsf{Hom}_K\big((L\otimes_K A)\times (L\otimes_K B), L\big)\\
&= \#\mathsf{Hom}_L(L^n\times L^m, L)\\
&= \#\mathsf{Hom}_L(L^{n+m}, L)\\
&= n + m\\
&= \#\mathsf{Hom}_K(A, L) + \#\mathsf{Hom}_K(B, L).
\end{aligned}
$$

This concludes the proof of the present lemma, but also that of Theorem 2.37   □

To conclude this chapter, let us observe that the Galois theorem 2.37 contains the classical Galois theorem 1.24. Indeed the contravariant equivalence of Theorem 2.37 implies in particular the existence of an isomorphism between the lattice of subobjects M

$$K \rightarrowtail M \rightarrowtail L$$

in $\mathsf{Split}[L : K]_f$ and the lattice of quotients $\mathsf{Hom}_K(M, L)$

$$\mathsf{Gal}[L : K] \cong \mathsf{Hom}_K(L, L) \twoheadrightarrow \mathsf{Hom}_K(M, L) \twoheadrightarrow \mathsf{Hom}_K(K, L) \cong \{*\}$$

in $\mathsf{Gal}[L : K]\text{-}\mathsf{Set}_f$. Propositions 2.21 and 2.32 show that this isomorphism is precisely the classical Galois isomorphism.

Chapter 3
Profinite Topological Spaces

Abstract The finite-dimensional Galois theorems of the previous chapters admit generalizations to arbitrary dimensions. This requires superposing topological structures on the algebraic ones. These topological aspects do not appear explicitly in the finite-dimensional cases, just because the topologies involved are then discrete. The aim of the present chapter is to develop the useful topological ingredients in view of proving infinite-dimensional Galois theorems. They will be obtained by a limit process from the finite discrete case. A topological space is *profinite* when it is compact Hausdorff and its topology admits a basis of closed open subsets, or equivalently, when it is a limit of finite discrete spaces. The Stone duality exhibits a contravariant equivalence of categories between the category of profinite spaces and that of Boolean algebras. This link will make it possible to combine algebraic and topological aspects in the infinite-dimensional Galois theory of fields, but also in the Galois theory of rings.

3.1 A Quick Review of Limits

The topological structures involved in infinite-dimensional Galois theory will be obtained by limit processes from finite discrete topologies.

Let us first recall a point of terminology.

Definition 3.1 A *small* category is one having a set of objects.

One should be aware that most categories encountered up to now: sets, groups, fields, algebras, and so on, are not small. For example:

Example 3.2 The category Set of sets is not small.

Proof Supposing that the category of sets is small, this would mean the existence of the set S of all sets. We could then consider the subset $S' \subseteq S$

$$S' = \{U \mid U \in S, \ U \notin U\}.$$

F. Borceux, *Galois Theories of Fields and Rings*, Coimbra Mathematical Texts 2,
https://doi.org/10.1007/978-3-031-58460-2_4

One has necessarily $S' \in S'$ or $S' \notin S'$, and exactly one of these two possibilities must hold. But none of these two possibilities can hold. Indeed, by definition of S', $S' \in S'$ would imply $S' \notin S'$ while $S' \notin S'$ would imply $S' \in S'$. $\square$

Definition 3.3 Let $F: X \longrightarrow \mathcal{A}$ be a functor.

1. A *cone* on F consists in giving an object $L \in \mathcal{A}$ and a family of morphisms $p_D: L \longrightarrow F(D)$, for all objects $D \in \mathcal{D}$, with the property that for each arrow $d: D \longrightarrow D'$ in $\mathcal{D}$, $F(d) \circ p_D = p_{D'}$.
2. This cone is a *limit* of the functor F when for every other cone $q_D: M \longrightarrow F(D)$, there exists a unique morphism $m: M \longrightarrow L$ such that for each D, $p_D \circ m = q_D$.

The category $\mathcal{A}$ is *complete* when the limit of every functor defined on a small category exists.

The dual notion, obtained by reversing the direction of every morphism, is that of *colimit*.

The uniqueness requirement in the definition of limit forces the limit, when it exists, to be unique up to an isomorphism.

The following particular cases of limits and colimits will be used intensively in this book. We use freely the notation of Definition 3.3.

Example 3.4 When the category $\mathcal{D}$ is discrete (each arrow is an identity), the functor F reduces to giving a family of objects and its possible limit is called the *product* of that family. In the category of sets, modules over a ring, or topological spaces, this is just the Cartesian product.
The dual notion is called the *coproduct* and denoted $\amalg$. In the category of sets, it is the disjoint union. In the category of R-modules, $A \amalg B = A \oplus B$.

Example 3.5 When the category $\mathcal{D}$ is reduced to two "parallel" arrows between two distinct objects (and of course the identities on these objects), the functor F reduces to giving a pair f, g

$$K \cdots \overset{k}{\cdots} > A \underset{g}{\overset{f}{\rightrightarrows}} B$$

of parallel arrows in $\mathcal{A}$; its possible limit is called the *equalizer* of these two arrows. The definition then reduces to giving a morphism $k: K \longrightarrow A$ such that $fk = gk$, while any other morphism with the same property factors uniquely through k. In the categories of sets, modules over a ring, or topological spaces, the equalizer K is the subset of elements $a \in A$ such that $f(a) = g(a)$ with the structure induced by that of A.
The dual notion is called the *coequalizer*. In the category of sets, it is the quotient of B by the equivalence relation generated by the pairs $(f(a), g(a))$. In the category of R-modules, it is the cokernel of $f - g$.

Example 3.6 When the category $\mathcal{D}$ is reduced to three objects D_0, D_1, D_2 and two arrows from D_1 and D_2 to D_0, the functor F reduces to giving two arrows f, g in $\mathcal{A}$ as in the diagram:

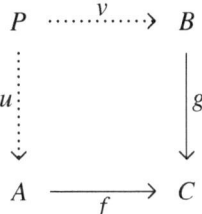

its possible limit is called the *pullback* of f and g. The definition reduces to giving two morphisms u, v such that $fu = gv$, while any other pair with the same property factors uniquely through it. In the case of sets, modules over a ring, or topological spaces, the pullback P is given by the pairs (a, b) in $A \times B$ such that $f(a) = g(b)$, with the structure induced by that of the product.

The dual notion is called the *pushout*.

The examples above show that the limits, in the category of R-modules, are computed just as in the category of sets. But as these examples also show, this is not at all the case for the corresponding colimits. However, in a special case of interest, it is.

We have already observed that a partially ordered set $(D, \leq)$ can be seen as a category whose objects are the elements of D and a single morphism is put from x to y when $x \leq y$. We shall often have to consider colimits defined on a filtered partially ordered set:

Definition 3.7 A partially ordered set $(D, \leq)$ is *filtered* when given x, y in D, there exists a z in D such that $x \leq z$ and $y \leq z$.[1]

Proposition 3.8 *Let $(D, \leq)$ be a filtered partially ordered set. The colimit of a functor $F: (D, \leq) \longrightarrow \mathsf{Set}$ to the category of sets admits the following description: on the disjoint union of the various $F(x)$, for all $x \in D$, one obtains an equivalence relation when declaring $a \in F(x)$ equivalent to $b \in F(y)$ if there exists a z in D, with $x \leq z$, $y \leq z$, such that a and b are mapped to the same element in $F(z)$. The limit of F is the quotient of the disjoint union by that equivalence relation.*

When the functor F takes values in the category of R-modules, its colimit admits as underlying set the corresponding filtered colimit of underlying sets.

Proof The case of sets is a routine exercise on equivalence relations. In the case of modules, consider first the limit L in Set of the corresponding underlying sets. Two elements in L are the equivalences classes of two elements $a \in F(x)$ and $b \in F(y)$, for some x, y in D. By the filteredness of D and the definition of the equivalence relation, there is no restriction in assuming that a, b lie in the same set $F(z)$. But $F(z)$ is an R-module, thus $a + b$ exists in $F(z)$ and its equivalence class defines the sum of the two given equivalence classes in L. The rest is routine.   □

Let us conclude this short review of limits with a very useful existence theorem.

---

[1] This notion of filteredness can be generalized to the case of an arbitrary category, not necessarily a partially ordered set, but we shall not need that generalization in this book.

Proposition 3.9 *A category is complete as soon as it admits equalizers and the product of every family of objects.*

Proof As we already know, products and equalizers are special cases of limits.

Conversely, if products and equalizers exist in $\mathcal{A}$, consider a functor $F\colon \mathcal{D} \longrightarrow \mathcal{A}$, with $\mathcal{D}$ small. For simplicity, given a morphism f in $\mathcal{D}$, we shall write D_f for its domain and C_f for its codomain, thus $f\colon D_f \longrightarrow C_f$. In the following diagram,

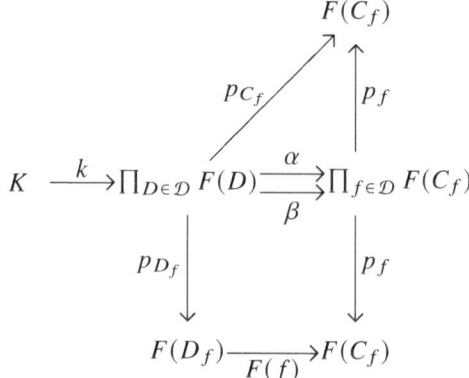

the notation p_* indicates the arrows of the corresponding product cone. The arrow α is the unique factorization of the cone p_{C_f} through the second product indexed by the arrows of $\mathcal{D}$, while β is the factorization through that same product of the cone $F(f) \circ p_{D_f}$. The morphism k is the equalizer of (α, β).

A routine chase on the diagram shows that $\left(p_D k\colon K \longrightarrow F(D)\right)_{D \in \mathcal{D}}$ is the expected limit. Indeed if $\left(q_D\colon M \longrightarrow F(D)\right)_{D \in \mathcal{D}}$ is another cone on F, by definition of the left-hand product, we have a unique factorization $m\colon M \longrightarrow \prod_{D \in \mathcal{D}} F(D)$ through the projections of this first product. The compatibility of the cone q_D with respect to an arrow f of $\mathcal{D}$ means $F(f)q_{D_f} = q_{C_f}$ and thus

$$p_f \beta m = F(f)p_{D_f}m = F_f q_{D_f} = q_{C_f} = p_{C_f}m = p_f \alpha m.$$

The uniqueness of a factorization through the right-hand product then implies $\alpha m = \beta m$, from which we get a further unique factorization $l\colon M \longrightarrow K$ of m through the equalizer k.   □

3.2 Totally Disconnected Spaces

Definition 3.10 A topological space X is *totally disconnected* when every two distinct elements $x \neq y$ of X are respectively contained in two disjoint clopens[2] U, V:

---

[2] "Clopen" meaning as usual a closed open subset.

$\forall x, y \in X \ \ x \neq y \Rightarrow \exists U, V$ clopens such that $x \in U$, $y \in V$, $U \cap V = \emptyset$.

Clearly, it is equivalent to require the existence of a clopen U such that $x \in U$ and $y \notin U$; it suffices then to put $V = \complement U$, the complement of U.

Proposition 3.11

1. *A totally disconnected space is Hausdorff.*
2. *Each subspace[3] of a totally disconnected space is still totally disconnected.*
3. *Each closed subspace of a compact totally disconnected space is still compact totally disconnected.*

Proof The first two statement are obvious and the third one holds because a closed subset of a compact Hausdorff space is compact.   □

Proposition 3.12 *The category of totally disconnected spaces is closed under limits in the category of topological spaces.*

Proof By Proposition 3.9, it suffices to treat separately the cases of products and equalizers.

Products in the category of topological spaces are just the topological products. And a topological product $\prod_{i \in I} X_i$ of totally disconnected spaces is still totally disconnected. Indeed, if $(x_i)_{i \in I} \neq (y_i)_{i \in I}$, there exists an index i_0 such that $x_{i_0} \neq y_{i_0}$. In the space X_{i_0}, there are thus disjoint clopens U_{i_0}, V_{i_0} such that $x_{i_0} \in U_{i_0}$ and $y_{i_0} \in V_{i_0}$. Putting $U_i = X_i = V_i$ for $i \neq i_0$, we obtain two disjoint clopens in the product

$$(x_i)_{i \in I} \in \prod_{i \in I} U_i \subseteq \prod_{i \in I} X_i, \quad (y_i)_{i \in I} \in \prod_{i \in I} V_i \subseteq \prod_{i \in I} X_i,$$

separating the given two elements, proving that $\prod_{i \in I} X_i$ is totally disconnected.

The case of equalizers follows at once from Proposition 3.11.1: an equalizer in Top is the equalizer in Set provided with the induced topology.   □

Corollary 3.13 *The category of compact totally disconnected spaces is closed under limits in the category of topological spaces.*

Proof A product of compact Hausdorff spaces is still compact Hausdorff (Tychonoff's theorem). Compact Hausdorff spaces are also closed under equalizers in the category of topological spaces. Indeed given $f, g \colon X \xrightarrow{\ \ \ } Y$ with X and Y compact Hausdorff, the diagonal of $Y \times Y$ is closed because Y is Hausdorff. Its inverse image along $(f, g) \colon X \longrightarrow Y \times Y$ is the equalizer of f and g, which is thus closed in the compact Hausdorff space X, thus is compact Hausdorff.

Combined with Proposition 3.12, these arguments complete the proof.   □

Let us observe an interesting property, and even a characterization, of compact totally disconnected spaces.

---

[3] By "subspace", we always mean a subset provided with the induced topology.

Corollary 3.14 *Let X be a compact Hausdorff space. The following conditions are equivalent:*

1. *X is totally disconnected;*
2. *the topology of X admits a base constituted of clopens.*

Proof $(2 \Rightarrow 1)$ is trivial. Conversely, a finite intersection of clopens of X is clopen. To prove that each open subset $U \subseteq X$ is a union of clopens, it suffices to prove that for each element $x \in U$, there exists a clopen V such that $x \in V \subseteq U$. Let us write $\complement$ to indicate the complement. Since U is open, $\complement U$ is closed, thus compact, in the compact space X. For each $y \notin U$, let us choose a clopen V_y such that $x \in V_y$ and $y \notin V_y$; thus $y \in \complement V_y$. The clopens $\complement V_y$ cover the compact subset $\complement U$, thus a finite number of them already cover $\complement U$. But

$$\complement U \subseteq \complement V_{y_1} \cup \cdots \cup \complement V_{y_n} = \complement \left(V_{y_1} \cap \cdots \cap V_{y_n} \right)$$

implies

$$x \in V_{y_1} \cap \cdots \cap V_{y_n} \subseteq U$$

with $V = V_{y_1} \cap \cdots \cap V_{y_n}$ a clopen. $\qquad\square$

Proposition 3.15 *In a compact totally disconnected space, two disjoint closed subsets are always contained in two disjoint clopens.*

Proof Let X be compact totally disconnected and $A, B \subseteq X$ two non-empty disjoint closed subsets (when one of them is empty, there is nothing to prove: X and $\emptyset$ do the job). In particular, A and B are compact.

For each fixed element $b \in B$ and each element $a \in A$, there exists a clopen U_a such that $a \in U_a$ and $b \notin U_a$. The union of these U_a covers the compact subset A, thus there exists a finite covering

$$A \subseteq U_{a_1} \cdots \cup U_{a_n} =_{\text{def}} U_b, \quad b \notin U_b.$$

We have thus already a clopen U_b which contains A but does not contain b; its complement $V_b = \complement U_b$ is a clopen which contains b and such that $A \cap V_b = \emptyset$.

The open subsets V_b cover the compact subset B, so there exists a finite covering

$$B \subseteq V_{b_1} \cup \cdots \cup V_{b_m} =_{\text{def}} W, \quad W \cap A = \emptyset.$$

We have obtained a clopen W which contains B, while $\complement W$ is a clopen which contains A. $\qquad\square$

Proposition 3.16 *In the category of compact totally disconnected spaces, performing the product with a space Z commutes with topological quotients.*

Proof Let X, Y, Z be compact totally disconnected spaces and $p \colon X \longrightarrow\!\!\!\!\!\rightarrow Y$ a topological quotient. We must prove that

$$\mathrm{id}_Z \times p \colon Z \times X \longrightarrow\!\!\!\!\!\rightarrow Z \times Y$$

is still a topological quotient.

Let us first prove that the quotient topology on $Z \times Y$ is compact totally disconnected; we shall prove next that it coincides with the product topology

The quotient topology on $Z \times Y$ is of course compact as a continuous image of the compact space $Z \times X$. Consider next $(z, y) \neq (z', y') \in Z \times Y$.

If $z \neq z'$, we can choose two disjoint clopens U, U' in Z with $z \in U \subseteq Z$, $z' \in U' \subseteq Z$. Then $U \times X$ and $U' \times X$ are disjoint clopens in $Z \times X$, which are saturated for the quotient. Their images in $Z \times Y$ are thus disjoint clopens separating (z, y) and (z', y').

Now if $z = z'$, then $y \neq y'$, and there exist disjoint clopens V, V' in Y, with $y \in V$ and $y' \in V'$. The disjoint subsets $Z \times V$ and $Z \times V'$ have inverse images $Z \times p^{-1}(V)$ and $Z \times p^{-1}(V')$ which are saturated clopens in $Z \times X$. Thus $Z \times V$ and $Z \times V'$ are disjoint clopens for the quotient topology and they separate (z, y) and (z, y').

It remains to prove that the quotient topology on $Z \times Y$ coincides with the product topology. The following composite

$$(Z \times X, \text{prod. top.}) \xrightarrow{(\mathrm{id}_Z \times R)} (Z \times Y, \text{quotient top.}) \xrightarrow{\mathrm{id}_{Z \times Y}} (Z \times Y, \text{prod. top.})$$

is the mapping $\mathrm{id}_Z \times p$, which is thus continuous between the two product spaces. By definition of a quotient topology, this proves that the mapping $\mathrm{id}_{Z \times Y}$ is continuous. But this is a continuous bijection between two compact Hausdorff spaces, thus a homeomorphism.   □

3.3 The Profinite Spaces

Definition 3.17 A topological space is profinite when it is compact totally disconnected.

The following proposition explains the terminology "profinite".

Theorem 3.18 *Let X be a topological space. The following conditions are equivalent:*

1. *X is a profinite space;*
2. *X is a cofiltered limit of finite discrete spaces;*
3. *X is a limit of finite discrete spaces;*
4. *X is the limit $X = \lim_{i \in I} X_i$ where*

   a. *the projections $p_i \colon X \longrightarrow X_i$ are all the topological quotients of X such that X_i is a finite discrete space;*
   b. *the set $(I, \leq)$ of indices is a cofiltered preordered set.*

Proof $(4 \Rightarrow 2)$ and $(2 \Rightarrow 3)$ are obvious. Each finite discrete space is compact totally disconnected. Thus $(3 \Rightarrow 1)$ holds by Corollary 3.13. It remains to prove $(1 \Rightarrow 4)$.

Let X be a compact, totally disconnected space. Consider the set $\mathcal{R}$, ordered by inclusion, of all the equivalence relations R on X, such that the topological quotient X/R is discrete finite. Let us write $[x]_R$ for the R-equivalence class of $x \in X$. Let us first prove that $\mathcal{R}$ is a cofiltered preordered set. Two relations $R, S \in \mathcal{R}$ correspond respectively to partitions $X = V_1 \cup \cdots \cup V_n$ and $X = W_1 \cup \cdots \cup W_m$. Let T be the equivalence relation corresponding to the partition

$$X = \bigcup_{1 \leq i \leq n,\ 1 \leq j \leq m,\ V_i \cap W_j \neq \emptyset} V_i \cap W_j.$$

Clearly $T \subseteq R$, $T \subseteq S$ and the quotient X/T is finite. Since X/R and X/S are discrete, all subsets V_i and W_j are clopens in X. Thus each $V_i \cap W_j$ is a clopen in X and therefore X/T is a discrete space, that is, $T \in \mathcal{R}$.

When $R \subseteq S$ in $\mathcal{R}$, we get a factorization $X/R \longrightarrow X/S$ between the quotients. We thus get a cofiltered diagram of finite discrete spaces. To conclude, it suffices to prove that $X \cong \lim_{R \in \mathcal{R}} X/R$. For this, let us consider the continuous factorization

$$\lambda \colon X \longrightarrow \lim_{R \in \mathcal{R}} X/R, \quad x \mapsto \big([x]_R\big)_{R \in \mathcal{R}};$$

we must prove that λ is a homeomorphism. On one hand, X is compact Hausdorff by assumption, thus $\lim_{R \in \mathcal{R}} X/R$ is compact Hausdorff by Corollary 3.13. Thus it suffices to prove that the continuous mapping λ is bijective. On the other hand $\lambda(X)$ is compact as a continuous image of a compact space; thus $\lambda(X)$ is closed in the Hausdorff space $\lim_{R \in \mathcal{R}} X/R$. To prove the bijectivity of λ, it thus suffices to prove that λ is injective and $\lambda(X)$ is dense in $\lim_{R \in \mathcal{R}} X/R$.

To prove that λ is injective, let us consider $x \neq y$ in X and $U \subseteq X$, a clopen such that $x \in U$ and $y \notin U$. Write X/U for the topological quotient admitting the clopens U and $\complement U$ as equivalence classes; it is thus the discrete space with two elements. The corresponding equivalence relation R is therefore in $\mathcal{R}$. Clearly, $[x]_R \neq [y]_R$ since $x \in U$ and $y \notin U$. Thus $\lambda(x) \neq \lambda(y)$.

To prove that $\lambda(X)$ is dense in $\lim_{R \in \mathcal{R}} X/R$, let us prove that $\lambda(X)$ contains an element of each fundamental open subset of $\lim_{R \in \mathcal{R}} X/R$. The fundamental open subsets are the finite intersections of the inverse images of the fundamental open subsets of the various factors X/R. Since each X/R is finite discrete, its fundamental open subsets are the singletons. The fundamental open subsets of $\lim_{R \in \mathcal{R}} X/R$ thus have the form

$$U = p_{R_1}^{-1}\big([x_1]_{R_1}\big) \cap \cdots \cap p_{R_n}^{-1}\big([x_n]_{R_n}\big),$$

where $x_1, \ldots, x_n \in X$, $R_1, \ldots, R_n \in \mathcal{R}$ and p_{R_i} are the projections of the limit. Let us choose an arbitrary element

$$\big([x_R]\big)_{R \in \mathcal{R}} \in U = p_{R_1}^{-1}\big([x_1]_{R_1}\big) \cap \cdots \cap p_{R_n}^{-1}\big([x_n]_{R_n}\big);$$

in particular, $[x_{R_i}]_{R_i} = [x_i]_{R_i}$ for each index $i = 1, \ldots, n$. Since $\mathcal{R}$ is cofiltered, we can choose $R_0 \in \mathcal{R}$ such that $R_0 \subseteq R_i$ for each $i = 1, \ldots, n$. And for each $i = 1, \ldots, n$, the existence of the inclusion $R_0 \subseteq R_i$, and the compatibility of the family $\left([x_R]_R\right)_{R\in\mathcal{R}}$, imply $[x_{R_i}]_{R_i} = [x_{R_0}]_{R_i}$. Thus

$$\lambda(x_{R_0}) = \left([x_{R_0}]_R\right)_{R\in\mathcal{R}} \in p_{R_1}^{-1}([x_{R_1}]_{R_1}) \cap \cdots \cap p_{R_n}^{-1}([x_{R_n}]_{R_n}) = U. \qquad \square$$

3.4 Boolean Algebras and Filters

To fix the notation and the terminology, let us recall some classical notions. We write $\complement x$ to indicate the complement of an element x of a Boolean algebra B:

$$x \wedge \complement x = 0, \quad x \vee \complement x = 1.$$

Definition 3.19 Let B be a Boolean algebra.

1. A *filter* in B is a subset $F \subseteq B$ such that

   (F1) $1 \in F$,
   (F2) $x \in F$ and $y \in F \Rightarrow x \wedge y \in F$,
   (F3) $x \in F$ and $x \le y \Rightarrow y \in F$.

2. The filter F is *proper* when

   (F4) $0 \notin F$.

3. An *ultrafilter* is a maximal element in the set of proper filters, ordered by inclusion.

   We write
   $$\uparrow x = \{y \in B \mid x \le y\}$$
   to indicate the *principal filter* generated by an element x of the Boolean algebra B.

Proposition 3.20 *Let $F \subseteq B$ be a proper filter in a Boolean algebra B. The following conditions are equivalent:*

1. *F is an ultrafilter;*
2. *$\forall x \in B \ \ x \in F$ or $\complement x \in F$;*
3. *$\forall x, y \in B \ \ x \vee y \in F \Rightarrow x \in F$ or $y \in F$;*
4. *there exists a homomorphism of Boolean algebras $f\colon B \longrightarrow \{0, 1\}$ such that $F = f^{-1}(1)$;*
5. *if G and H are filters,*

$$G \cap H \subseteq F \Rightarrow G \subseteq F \text{ or } H \subseteq F.$$

Proof $(1 \Rightarrow 2)$. If $x \notin F$, then $G = \{z \mid x \vee z \in F\}$ is a proper filter containing F. Therefore, $G = F$ and $\complement x \in G = F$.

$(2 \Rightarrow 3)$. If $x \vee y \in F$ and $x \notin F$, we have $\complement x \in F$. Thus $y \wedge \complement x = (x \vee y) \wedge \complement x \in F$. This implies $y \in F$.

$(3 \Rightarrow 4)$. Let us put $f(x) = 1$ if $x \in F$ and $f(x) = 0$ otherwise. The mapping f preserves $\wedge$, $\leq$, 1 because F is a filter. It preserves 0 as well because F is proper. Condition 3 implies that f preserves $\vee$. The preservation of $\complement$ follows from the preservation of 0, 1, $\wedge$ and $\vee$.

$(4 \Rightarrow 1)$. Condition 2 is satisfied because f preserves $\complement$. This implies that when G is a filter which contains F, for each element $x \in G \setminus F$ we have $\complement x \in F$. Thus $0 = x \wedge \complement x \in G$ and this proves that $G = B$.

$(2 \Rightarrow 5)$. If $G \nsubseteq F$ and $H \nsubseteq F$, consider $x \in G \setminus F$ and $y \in H \setminus F$. Then $\complement x \in F$, $\complement y \in F$; this implies $\complement (x \vee y) = \complement x \wedge \complement y \in F$. So $x \vee y \in (G \cap H) \setminus F$.

$(5 \Rightarrow 3)$. If $x \vee y \in F$, then $\uparrow x \wedge \uparrow y \subseteq F$. Thus $\uparrow x \subseteq F$ or $\uparrow y \subseteq F$.   □

Proposition 3.21 *In a Boolean algebra B,*

1. *each proper filter is contained in an ultrafilter;*
2. *each non-zero element belongs to an ultrafilter;*
3. *$x \nleq y \Rightarrow \exists F$ ultrafilter, with $x \in F$, $y \notin F$;*
4. *each filter is the intersection of the ultrafilters which contain it.*

Proof The proper filters containing a given filter F constitute an inductive preordered set: that is, the union of a totally ordered family of such filters is again such a filter, containing F but not containing 0. Zorn's lemma[4] then implies condition 1.

It suffices to apply condition 1 to the principal filter $\uparrow x$ to get condition 2.

Condition 2 applied to $x \wedge \complement y$ yields condition 3. This also implies condition 4 in the case of a proper filter.

Condition 4 in the case of the non-proper filter $B \subseteq B$ is the trivial observation that the intersection of the empty family of ultrafilters is B itself.   □

3.5 The Spectrum of a Boolean Algebra

The spectrum of a Boolean algebra is a topological space canonically associated with the Boolean algebra. We shall see in the next section that this topological space completely characterizes the given Boolean algebra.

Proposition 3.22 *Let B be a Boolean algebra and* Spec(B) *the set of its ultrafilters. For each filter H of B, let us consider*

$$O_H = \{F \in \text{Spec}(B) \mid H \nsubseteq F\}.$$

Writing $\wp(B)$ to indicate the Boolean algebra of subsets of B,

1. *the mapping*

$$O : \text{Filters}(B) \longrightarrow \wp(B), \quad H \mapsto O_H$$

is a full inclusion of preordered sets;

---

[4] Zorn's lemma, which is equivalent to the axiom of choice, asserts that in a partially ordered set, if every chain has an upper bound, then there exists at least one maximal element.

2. *the subsets* $O_H \subseteq \mathsf{Spec}(B)$ *constitute a topology* $\mathcal{T}$ *on the set* $\mathsf{Spec}(B)$.

Proof The mapping O is injective by condition 4 in Proposition 3.21. Moreover if $H \subseteq H'$ are filters, we get at once $O_H \subseteq O_{H'}$. Conversely $O_H \subseteq O_{H'}$, for arbitrary filters H and H', implies $H \subseteq H'$, again by condition 4 in Proposition 3.21. Thus O is a full inclusion of preordered sets

Clearly, $\emptyset = O_{\{1\}}$ and $\mathsf{Spec}(B) = O_B$. Moreover, if H and H' are filters, $O_H \cap O_{H'} = O_{H \cap H'}$ by condition 5 in Proposition 3.20. Finally if $(H_i)_{i \in I}$ is a family of filters and H is the filter generated by their union, the equality $\bigcup_{i \in I} O_i = O_H$ is trivial. $\qquad\square$

Definition 3.23 Let B be a Boolean algebra. The topological space $\mathsf{Spec}(B)$ defined in Proposition 3.22 is called the *spectrum* of the Boolean algebra B.

Proposition 3.24 *Let* B *be a Boolean algebra. For each element* $b \in B$, *the open subset*

$$O_b \equiv O_{\uparrow b} = \{ F \in \mathsf{Spec}(B) | b \notin F \}$$

is closed. These clopens O_b *constitute a basis of the topology of the spectrum* $\mathsf{Spec}(B)$ *and satisfy the following conditions:*

1. $b \le b' \Rightarrow O_b \supseteq O_{b'}$,
2. $b \ne b' \Rightarrow O_b \ne O_{b'}$,
3. $O_{b \wedge b'} = O_b \cup O_{b'}$,
4. $O_{b \vee b'} = O_b \cap O_{b'}$,
5. $O_1 = \emptyset, \quad O_0 = B$,
6. $O_{\complement b} = \complement O_b$,

for each pair of elements $b, b' \in B$.

Proof The six conditions are direct consequences of Propositions 3.20 and 3.21. Condition 6 implies that O_b is closed. For each filter $H \subseteq B$, $H = \bigcup_{b \in H} \uparrow b$. Thus $O_H = \bigcup_{b \in H} O_{\uparrow b}$ (see Proposition 3.22). And by condition 4, $O_{\uparrow b} \cap O_{\uparrow b'} = O_{b \vee b'}$. $\square$

Proposition 3.25 *The spectrum of a Boolean algebra is a profinite space.*

Proof If $F \ne F'$ are distinct ultrafilters, the maximality of F implies $F \nsubseteq F'$. Choosing $b \in F \setminus F'$, we get $F' \in O_b$ and $F \notin O_b$. By Proposition 3.24, $\mathsf{Spec}(B)$ is totally disconnected.

It remains to prove that $\mathsf{Spec}(B)$ is compact. Let $\mathsf{Spec}(B) = \bigcup_{i \in I} O_{b_i}$ be a covering by basic open subsets (see Proposition 3.24). We have

$$\forall F \in \mathsf{Spec}(B) \quad \exists i \in I \quad F \in O_{b_i}$$

or, equivalently,

$$\forall F \in \mathsf{Spec}(B) \quad \exists i \in I \quad b_i \notin F.$$

Consider now

$$G = \{ y \in B \mid \exists i_1, \dots, i_n \in I \quad b_{i_1} \wedge \cdots \wedge b_{i_n} \le y \}.$$

We conclude by a reduction *ad absurdum*. If no finite subcovering of the covering exists, for each sequence $b_{i_1}, \ldots, b_{i_n}$, there exists an ultrafilter $F_{i_1,\ldots,i_n}$ such that

$$F_{i_1,\ldots,i_n} \notin O_{b_{i_1}} \cup \cdots \cup O_{b_{i_n}} = O_{b_{i_1} \wedge \cdots \wedge b_{i_n}},$$

that is, $b_{i_1} \wedge \cdots \wedge b_{i_n} \in F_{i_1,\ldots,i_n}$. In particular $b_{i_1} \wedge \cdots \wedge b_{i_n} \neq 0$. This proves that G, which is trivially a filter, is in fact a proper filter. By Proposition 3.21, this proper filter G is contained in some ultrafilter F. But G, and thus F, contain all the elements $b_i, i \in I$. This implies $F \notin O_{b_i}$, for each $i \in I$: this is a contradiction.   □

Corollary 3.26 *A subset $U \subseteq \mathsf{Spec}(B)$ of a Boolean algebra B is clopen precisely when it has the form O_b for some element $b \in B$.*

Proof We know already that each O_b is a clopen (see Proposition 3.24). Conversely if U is a clopen, $U = \bigcup_{i \in I} O_{b_i}$, for elements $b_i \in B$ (see Proposition 3.24). But U is compact because it is closed in the compact space $\mathsf{Spec}(B)$ (see Proposition 3.25); thus there exists a finite subcovering

$$U = O_{b_1} \cup \cdots \cup O_{b_n} = O_{b_1 \wedge \cdots \wedge b_n}.$$ □

3.6 Stone Duality

This section presents the Stone duality theorem: the category of Boolean algebras is dual to the category of profinite spaces.

Proposition 3.27 *Each Boolean algebra B is isomorphic to the Boolean algebra of clopens of its spectrum $\mathsf{Spec}(B)$.*

Proof By Propositions 3.26 and 3.24

$$B \longrightarrow \{U \subseteq \mathsf{Spec}(B) | U \text{ clopen}\}, \quad b \mapsto O_{C_b}$$

is an isomorphism of Boolean algebras.   □

Proposition 3.28 *Each profinite space X is the spectrum of the Boolean algebra $\mathsf{Clopen}(X)$ of its clopens.*

Proof The clopens of every topological space constitute trivially a Boolean algebra. If $x \in X$, let us put

$$F_x = \{U \subseteq X \mid U \text{ clopen}, x \in U\}.$$

Each F_x is a proper filter in the Boolean algebra of clopens of X. It is in fact an ultrafilter because

$$\forall U \in \mathsf{Clopen}(X) \quad x \in U \text{ or } x \notin U$$

(see Proposition 3.20). Let us consider the mapping

$$\varphi \colon X \longrightarrow \mathsf{Spec}\big(\mathsf{Clopen}(X)\big), \quad x \mapsto F_x;$$

we shall now prove that φ is a homeomorphism.

If $x \neq y$, by Theorem 3.18 there exists a clopen U such that $x \in U$ and $y \in \complement U$. Thus $U \in F_x$ and $U \notin F_y$. This implies $F_x \neq F_y$, thus φ is injective.

If $\mathcal{F} \subseteq \mathsf{Clopen}(X)$ is an ultrafilter, let us consider

$$V = \bigcap \{U \subseteq X \mid U \in \mathcal{F}\}.$$

The elements $U \in \mathcal{F}$ are in particular closed in X. Since $\mathcal{F}$ is a proper filter, a finite intersection of elements in $\mathcal{F}$ is never empty; and since X is compact, this implies that V is not empty. We choose then $x \in V$. The proper filter F_x contains $\mathcal{F}$, thus $\mathcal{F} = F_x = \varphi(x)$ by maximality of $\mathcal{F}$. This proves that φ is surjective.

Since φ is a bijection between compact Hausdorff spaces, it suffices to prove that φ is continuous, to conclude that it is a homeomorphism. Applying Proposition 3.24, let us consider a basic open subset of $\mathsf{Spec}\big(\mathsf{Clopen}(X)\big)$; this basic open subset has the form O_W for some element $W \in \mathsf{Clopen}(X)$. Thus

$$\varphi^{-1}(O_W) = \{x \in X \mid W \notin F_x\} = \{x \in X \mid x \notin W\} = \complement W,$$

where $\complement W$ is open in X because W is closed. $\qquad\qquad\square$

Theorem 3.29 (Stone duality) *There exists a contravariant equivalence of categories, where*

$$\mathsf{Boole} \underset{\mathsf{Clopen}}{\overset{\mathsf{Spec}}{\rightleftarrows}} \mathsf{Prof}$$

1. Boole *is the category of Boolean algebras;*
2. Prof *is the category of profinite spaces;*
3. $\mathsf{Spec}(B)$ *is the spectrum of the Boolean algebra B;*
4. $\mathsf{Clopen}(X)$ *is the Boolean algebra of clopens of the profinite space X.*

Proof If $f \colon B \longrightarrow B'$ is a homomorphism of Boolean algebras, composition with f

$$\mathsf{Hom}\big(B', \{0, 1\}\big) \longrightarrow \mathsf{Hom}\big(B, \{0, 1\}\big), \quad h \mapsto hf$$

yields, by Proposition 3.20, a mapping

$$\mathsf{Spec}(f) \colon \mathsf{Spec}(B') \longrightarrow \mathsf{Spec}(B).$$

This mapping is continuous because, when $b \in B$,

$$\begin{aligned}
\mathsf{Spec}(f)^{-1}(O_b) &= \big\{F \in \mathsf{Spec}(B') \big| b \notin \mathsf{Spec}(f)(F)\big\} \\
&= \big\{F \in \mathsf{Spec}(B') \big| b \notin f^{-1}(F)\big\} \\
&= \big\{F \in \mathsf{Spec}(B') \big| f(b) \notin F\big\} \\
&= O_{f(b)}.
\end{aligned}$$

This defines the functor Spec.

Now if $g: X \longrightarrow X'$ is a continuous mapping between profinite spaces,

$$g^{-1}: \mathsf{Clopen}(X') \longrightarrow \mathsf{Clopen}(X)$$

is a morphism of Boolean algebras, and this defines the functor Clopen.

Propositions 3.27 and 3.28 imply already that Spec and Clopen are mutually inverse on the objects. If now $f: B \longrightarrow B'$ is a homomorphism of Boolean algebras,

$$(\mathsf{Clopen} \circ \mathsf{Spec})(f)(O_b) = \mathsf{Spec}(f)^{-1}(O_b) = O_{f(b)},$$

thus $\mathsf{Clopen} \circ \mathsf{Spec} \cong \mathsf{id}$. And if $g: X \longrightarrow X'$ is a continuous mapping between profinite spaces and $F_x \subseteq X$ is the ultrafilter in $\mathsf{Clopen}(X)$ which corresponds to $x \in X$ by the isomorphism $X \cong \mathsf{Spec}(\mathsf{Clopen}(X))$, $\mathsf{Spec} \circ \mathsf{Clopen} \cong \mathsf{id}$ holds as well because

$$\begin{aligned}
(\mathsf{Spec} \circ \mathsf{Clopen})(g)(F_x) &= \left\{ W \in \mathsf{Clopen}(X') \middle| g^{-1}(W) \in F_x \right\} \\
&= \left\{ W \in \mathsf{Clopen}(X') \middle| x \in g^{-1}(W) \right\} \\
&= \left\{ W \in \mathsf{Clopen}(X') \middle| g(x) \in W \right\} \\
&= F_{g(x)}. \qquad \qquad \square
\end{aligned}$$

3.7 Finite Boolean Algebras

Let us conclude this chapter with some specific properties of finite Boolean algebras.

Proposition 3.30 *Each finite Boolean algebra is isomorphic to the Boolean algebra of subsets of a finite set.*

Proof The spectrum of a Boolean algebra is (compact) Hausdorff by Proposition 3.25. In particular each point is closed, thus each finite subset is closed. But the spectrum of a finite Boolean algebra is trivially finite, thus each of its subsets is closed and the spectrum is discrete. The clopens are thus all the subsets of the spectrum and Proposition 3.27 forces the conclusion. $\square$

Proposition 3.31 *Each Boolean algebra is the filtered union of its finite Boolean subalgebras.*

Proof Via the Stone duality (see Theorem 3.29), the finite discrete profinite spaces correspond to their Boolean algebras of subsets, that is, by Proposition 3.30, to the finite Boolean algebras. Thus this Proposition 3.31 is exactly Theorem 3.18 transposed via the Stone duality 3.29. $\square$

Proposition 3.32 *Let A be a finite Boolean algebra and $B = \bigcup_{i \in I} B_i$ a Boolean algebra presented as a filtered union of finite Boolean subalgebras B_i. The canonical factorization*

$$\lambda: \operatorname*{colim}_{i \in I} \operatorname{Hom}(A, B_i) \longrightarrow \operatorname{Hom}\left(A, \bigcup_{i \in I} B_i\right) = \operatorname{Hom}(A, B)$$

is bijective, where Hom *indicates the set of homomorphisms of Boolean algebras.*

Proof It is immediate that a set-theoretical filtered union of Boolean subalgebras remains a Boolean subalgebra. Indeed if x and y are elements in the union, there are indices i, j such that $x \in B_i$ and $y \in B_j$. But by filteredness, B_i and B_j are contained in some B_k; thus x, y are in B_k and therefore also $x \wedge y, x \wedge y, \complement x, \complement y$. So all these elements are in the union.

Let us write $\beta_i: B_i \hookrightarrow B$ for the canonical inclusions and $\beta_{i,j}: B_i \hookrightarrow B_j$ for the possible inclusion of B_i in B_j. Just as in Proposition 3.8, the colimit in the statement is defined in the following way. One considers first the set of all homomorphisms of Boolean algebras $f_i: A \longrightarrow B_i$, for all indices $i \in I$, and one performs the quotient identifying $f_i: A \longrightarrow B_i$ and $f_j: A \longrightarrow B_j$ when there exists a B_k containing both B_i and B_j while $\beta_{ik} \circ f_i = \beta_{jk} \circ f_j$. Of course this is equivalent to saying that $\beta_i \circ f_i = \beta_j \circ f_j$ as homomorphisms from A to B. One defines of course $\lambda([f_i]) = \beta_i \circ f_i$.

In particular, with the notation above, $\lambda([f_i]) = \lambda([f_j])$ means that $f_i: A \longrightarrow A_i$ and $f_j: A \longrightarrow A_j$ are equal when viewed as morphisms $A \longrightarrow B$. In particular for each element $a \in A$, $f_i(a) = f_j(a)$ in B, thus in some B_{k_a}. Since A is finite, we can choose an index k such that $k_a \leq k$ for each a, and thus f_i and f_j are equal as morphisms from A to B_k. They are thus equivalent in the colimit. This proves that λ is injective.

Next start with a homomorphism $h: A \longrightarrow B$ of Boolean algebras. By Proposition 3.30, A is isomorphic to the Boolean algebra $\wp(X)$ of subsets of a finite set X. Since $B = \bigcup_{i \in I} B_i$, for each element $x \in X$ there exists an index i_x such that $h(\{x\}) \in B_{i_x}$. Since X is finite, by filteredness, we can choose a single index i such that $i_x \leq i$ and $h(\{x\}) \in B_i$ for each $x \in X$. Thus h admits a corestriction $h': A \cong \wp(X) \longrightarrow B_i$ and $h = \lambda([h'])$. $\square$

Proposition 3.32 is a special instance of the so-called theory of *finitely presentable objects* in a category: an object A such that the functor $\operatorname{Hom}(A, -)$ preserves filtered colimits.

Chapter 4
The Galois Theorems in Arbitrary Dimension

Convention. *In this chapter, all fields are commutative; all algebras are commutative with unit.*

Abstract In Chapter 3 we pointed out an equivalent definition of a finite-dimensional Galois extension in terms of a tensor product, instead of polynomials over a field. This is a first step towards a Galois theory for rings, where the polynomial approach fails to work. The present chapter develops a second important step in the same direction: getting rid of the notion of dimension, which does not naturally make sense in the case of rings. We thus generalize both the classical Galois theorem and the Grothendieck Galois theorem, from the finite-dimensional case to the case of a Galois extension of fields $K \subseteq L$ of arbitrary dimension. This requires introducing, on the corresponding Galois group $\mathsf{Gal}[L : K]$, a profinite topology, as studied in Chapter 3. The classical Galois theorem then exhibits a bijection between the intermediate field extensions and the closed subgroups of the Galois group. In the Grothendieck approach in terms of split algebras, the $\mathsf{Gal}[L : K]$-sets are now provided with a profinite topology making the group actions continuous.

4.1 Arbitrary Galois Extensions

Let $K \subseteq L$ be a Galois extension of fields. Let us investigate the subextensions $K \subseteq M \subseteq L$ where $K \subseteq M$ is finite-dimensional. We call this for short a *finite-dimensional subextension*.

Proposition 4.1 *Let $K \subseteq L$ be a Galois extension of fields. Consider an element $l \in L$ and its minimal polynomial $p(X) \in K[X]$, with roots $l_1, \ldots, l_n$ in L. Then $K \subseteq K(l_1, \ldots, l_n) \subseteq L$ is a finite-dimensional Galois subextension.*

Proof The subfield $K \subseteq K(l_1, \ldots, l_n) \subseteq L$ generated by $l_1, \ldots, l_n$ is finite-dimensional over K and is also the K-subalgebra of L generated by the elements $l_1, \ldots, l_n$ (see Proposition 2.19). Each element l_i admits $p(X)$ as minimal polynomial, thus by Proposition 2.12

F. Borceux, *Galois Theories of Fields and Rings*, Coimbra Mathematical Texts 2, https://doi.org/10.1007/978-3-031-58460-2_5

$$\#\mathrm{Hom}_K \left(K(l_i), K(l_1, \ldots, l_n) \right)$$

$$= \#\mathrm{Hom}_K \left(\frac{K[X]}{\langle p(X) \rangle}, K(l_1, \ldots, l_n) \right)$$

$$= \text{number of roots of } p(X) \text{ in } K(l_1, \ldots, l_n)$$

$$= \text{degree of } p(X)$$

$$= \dim {}_K K(l_i).$$

This proves that each algebra $K(l_i)$ is split by the extension $K \subseteq K(l_1, \ldots, l_n)$ (see Theorem 2.27). But $K(l_1, \ldots, l_n)$ is the union, as K-subalgebras, of the K-subalgebras $K(l_i)$. Thus $K(l_1, \ldots, l_n)$ is split by the extension $K \subseteq K(l_1, \ldots, l_n)$ (see Lemma 2.29) and $K \subseteq K(l_1, \ldots, l_n)$ is therefore a Galois extension (see Proposition 2.25). $\qquad\square$

Proposition 4.2 *Let $K \subseteq L$ be a Galois extension of fields and $K \subseteq M \subseteq L$ a finite-dimensional subextension. Every K-automorphism of fields $f \colon M \longrightarrow M$ extends as a K-automorphism $g \colon L \longrightarrow L$.*

Proof By Proposition 1.10, it suffices to exhibit a K-endomorphism $g \colon L \longrightarrow L$ extending f: it will automatically be a K-automorphism. Let us perform a transfinite induction on the ordinals, using Lemma 1.15, to construct a sequence $f_\alpha \colon N_\alpha \longrightarrow L$ of extensions of f.

- $N_0 = M$, $f_0 = f \colon N_0 \longrightarrow L$.
- If $N_\alpha = L$, $f_\alpha \colon L \longrightarrow L$ is an extension of f and the proof is done. Otherwise, choose $l \in L$, $l \notin N_\alpha$; let us put $N_{\alpha+1} = N_\alpha(l)$ and $f_{\alpha+1} = \widetilde{f_\alpha}$, where $\widetilde{f_\alpha}$ is an extension given by Lemma 1.15.
- If α is a limit ordinal, let us put $N_\alpha = \bigcup_{\beta < \alpha} N_\beta$, with f_α equal to f_β on N_β, for each $\beta < \alpha$.

The process will stop at some $N_\alpha = L$, after a number of steps less than or equal to the cardinal of L. $\qquad\square$

Proposition 4.3 *Let $K \subseteq L$ be a Galois extension of fields. The field L is the filtered union of all the finite-dimensional Galois subextensions $K \subseteq M \subseteq L$.*

Proof If $l \in L$ admits the minimal polynomial $p(X) \in K[X]$, with roots $l_1, \ldots, l_n$ in L, Proposition 4.1 implies

$$l \in K(l_1, \ldots, l_n) \subseteq L,$$

where $K \subseteq K(l_1, \ldots, l_n)$ is a finite-dimensional Galois subextension. Thus the field L is indeed the set-theoretical union of all its finite-dimensional Galois subextensions.

It remains to prove that this union is filtered. For this let us consider two finite-dimensional Galois subextensions

$$K \subseteq M_1 \subseteq L, \quad K \subseteq M_2 \subseteq L.$$

The K-subalgebra $M_3 \subseteq L$ generated by M_1 and M_2 is still finite-dimensional over K; it is thus algebraic over K (see Proposition 2.15). It is a field by Corollary 2.21. Since $K \subseteq M_1$ is a Galois extension, the minimal polynomial of an element $l \in M_1$ admits a decomposition into distinct linear factors in $M_1[X]$, thus *a fortiori* in $M_3[X]$; the same argument holds for M_2. This proves that M_1 and M_2 are split by the field extension $K \subseteq M_3$. But M_3 is the union of M_1 and M_2 as K-subalgebras; Lemma 2.29 implies therefore that M_3 is split by the extension $K \subseteq M_3$, thus $K \subseteq M_3$ is a Galois extension (see Proposition 2.25). □

4.2 The Topological Galois Group

The topological Galois group of a field extension $K \subseteq L$ is the classical Galois group (see Definition 1.11) $\mathsf{Gal}[L : K]$ provided with a profinite topology which, in the finite-dimensional case, turns out to be the discrete one.

Proposition 4.4 *Let $K \subseteq L$ be a field extension. Then*

$$\mathsf{Gal}[L : K] = \lim_M \mathsf{Gal}[M : K]$$

where

- $K \subseteq M \subseteq L$ *are all the finite-dimensional intermediate Galois extensions;*
- *when $M \subseteq M'$, $\mathsf{Gal}[M' : K] \longrightarrow \mathsf{Gal}[M : K]$ is the restriction mapping;*
- *the limit is cofiltered;*
- *each projection of the limit is surjective;*
- *the kernel of the projection $\mathsf{Gal}[L : K] \longrightarrow \mathsf{Gal}[M : K]$ is the subgroup $\mathsf{Gal}[L : M]$.*

Proof The projections $p_M \colon \mathsf{Gal}[L : K] \longrightarrow \mathsf{Gal}[M : K]$ are the corresponding restrictions (see Proposition 1.17) and constitute a cone over the diagram of the statement. To prove that this cone is a limit cone, let us choose a compatible family $f_M \in \mathsf{Gal}[M : K]$. Since $L = \bigcup_M M$ is a filtered union (see Proposition 1.16), but also a union computed as in the Set case, the K-homomorphisms $f_M \colon M \longrightarrow L$ induce a K-homomorphism $f \colon L \longrightarrow L$, that is, an element $f \in \mathsf{Gal}[L : K]$ (see Proposition 1.10).

The diagram is cofiltered by Proposition 4.3. The projections are surjective by Proposition 1.16. The restriction to M of a K-homomorphism $f \in \mathsf{Gal}[L : K]$ is the identity on M precisely when f is an M-homomorphism, that is, when $f \in \mathsf{Gal}[L : M]$. □

Let us recall that a topological group is a group provided with a topology which makes continuous the group operations (multiplication, inverse). A limit of topological groups is just their limit as sets, provided with both the structure of a group and of a topological space induced by that of the Cartesian product (see Proposition 3.9).

Definition 4.5 Let $K \subseteq L$ be a Galois extension of fields. The topological Galois group of this extension is the limit

$$\mathsf{Gal}[L : K] = \lim_M \mathsf{Gal}[M : K]$$

in the category of topological groups, where $K \subseteq M \subseteq L$ runs through the finite-dimensional intermediate field extensions and each finite group $\mathsf{Gal}[M : K]$ is provided with the discrete topology.

Proposition 4.6

1. *The topological Galois group of a Galois extension of fields is a profinite space.*
2. *The limit in Definition 4.5 is cofiltered.*
3. *Each projection in Definition 4.5 is a topological quotient.*

Proof With the notation of Definition 4.5, the topological group $\mathsf{Gal}[L : K]$ is profinite, by Theorem 3.18 and Corollary 3.13. The limit is cofiltered by Proposition 4.3.

The quotient topology on each term $\mathsf{Gal}[M : K]$ is the finest topology such that the projection is continuous. But the projection is continuous for the discrete topology on $\mathsf{Gal}[M : K]$: thus the discrete topology is the quotient topology.   □

Proposition 4.7 *The topological Galois group of a finite-dimensional Galois extension of fields is a discrete topological space.*

Proof With the notation of Definition 4.5, the Galois group $\mathsf{Gal}[L : K]$ with the discrete topology is an initial object in the diagram which defines its topology; it is thus the limit of the diagram.   □

Let us now study more precisely the topology of the Galois group. It is probably useful to recall that in a topological group, it suffices to know the neighborhoods of the unit element 1 in order to know the whole topology of the group. Indeed each mapping

$$g \cdot (-) : G \longrightarrow G, \quad g' \mapsto g \cdot g'$$

is a homeomorphism with inverse $g^{-1} \cdot (-)$. This homeomorphism maps 1 to g, and thus transforms the neighborhoods of 1 into the neighborhoods of g.

Proposition 4.8 *Let $K \subseteq L$ be a Galois extension of fields. The subgroups $\mathsf{Gal}[L : M] \subseteq \mathsf{Gal}[L : K]$, for all intermediate finite-dimensional Galois field extensions $K \subseteq M \subseteq L$, constitute a fundamental system of closed-open neighborhoods of the unit $\mathsf{id}_L \in \mathsf{Gal}[L : K]$ of the Galois group.*

Proof We know already (see Proposition 4.4) that $\mathsf{Gal}[L : M] = p_M^{-1}(\{\mathsf{id}_M\})$, with $\{\mathsf{id}_M\}$ clopen in the discrete space $\mathsf{Gal}[M : K]$. Thus $\mathsf{Gal}[L : M]$ is a closed-open neighborhood of id_L in $\mathsf{Gal}[L : K]$.

Conversely, since each term of the limit in Definition 4.5 is discrete, a fundamental neighborhood of id_L has the form

$$V = p_{M_1}^{-1}(\{1_{M_1}\}) \cap \cdots \cap p_{M_n}^{-1}(\{1_{M_n}\}),$$

where $M_1, \ldots, M_n$ are finite-dimensional Galois subextensions. Since each $\{\mathrm{id}_{M_i}\}$ is clopen in $\mathrm{Gal}[M_i, K]$, the neighborhood V is clopen in $\mathrm{Gal}[L : K]$. Let us further observe that

$$f \in V \Leftrightarrow \forall i = 1, \ldots, n \;\; f|_{M_i} = \mathrm{id}_{M_i} \Leftrightarrow f|_M = \mathrm{id}_M$$

where M is the subextension generated by $M_1, \ldots, M_n$, that is, $V = \mathrm{Gal}[L : M]$. We know that $K \subseteq M$ is also a finite-dimensional Galois extension (this is precisely the proof of Proposition 4.3). $\qquad\square$

Proposition 4.9 *Let $K \subseteq L$ be a Galois field extension. The topology of the Galois group $\mathrm{Gal}[L : K]$ is the initial topology[1] for all the mappings*

$$\mathrm{ev}_l \colon \mathrm{Gal}[L : K] \longrightarrow L, \;\; f \mapsto f(l)$$

where $l \in L$ and L is provided with the discrete topology.

Proof First, let us prove that the mapping ev_l is continuous for the topology defined in Definition 4.5. Since L is provided with the discrete topology, it suffices to prove that for each element $l_0 \in L$

$$\mathrm{ev}_l^{-1}(\{l_0\}) = \{f \in \mathrm{Gal}[L : K] \,|\, f(l) = l_0\}$$

is open in $\mathrm{Gal}[L : K]$. Of course if this subset is empty, it is open. Otherwise, let us choose $f_0 \in \mathrm{ev}_l^{-1}(\{l_0\})$. By Proposition 4.3, there exists a finite-dimensional Galois subextension $K \subseteq M \subseteq L$ which contains l and l_0. Let $p(X) \in K[X]$ be the minimal polynomial of l. Since f_0 is a K-homomorphism,

$$p(l_0) = p(f_0(l)) = f_0(p(l)) = f_0(0) = 0$$

and l_0 is a root of $p(X)$. By Proposition 1.20, there exists a K-automorphism $g \colon M \longrightarrow M$ such that $g(l) = l_0$. By Proposition 4.2, there is further a K-automorphism $\overline{g} \colon L \longrightarrow L$ such that $\overline{g}\,|_M = g$.

$$
\begin{aligned}
\mathrm{ev}_l^{-1}(\{l_0\}) &= \{f \in \mathrm{Gal}[L : K] \,|\, f(l) = l_0\} \\
&= \{f \in \mathrm{Gal}[L : K] \,|\, f(l) = g(l)\} \\
&= \{f \in \mathrm{Gal}[L : K] \,|\, f(l) = \overline{g}(l)\} \\
&= \{f \in \mathrm{Gal}[L : K] \,|\, (\overline{g}^{-1} \circ f)|_M(l) = l\}.
\end{aligned}
$$

Let us consider the pullback

---

[1] Given a family $(f_i \colon A \longrightarrow B_i)_{i \in I}$ of mappings, with A a set and each B_i a topological space, the corresponding initial topology on A is the coarsest topology containing all the $f_i^{-1}(U)$ for all indices $i \in I$ and all open subsets $U \subseteq B_i$.

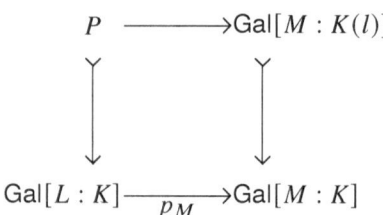

We have
$$P = \{h \in \text{Gal}[L : K] \,\big|\, h|_M(l) = l\}$$

thus
$$\text{ev}_l^{-1}(\{l_0\}) = \{f \in \text{Gal}[L : K] \,\big|\, \overline{g}^{-1} \circ f \in P\} = \{\overline{g} \circ h \,\big|\, h \in P\}.$$

This proves that $\text{ev}_l^{-1}(\{l_0\})$ is the image of the subset $P \subseteq \text{Gal}[L : K]$ under the homeomorphism

$$\text{Gal}[L : K] \longrightarrow \text{Gal}[L : K], \quad h \mapsto \overline{g} \circ h.$$

Since $\text{Gal}[M : K]$ has the discrete topology, $\text{Gal}[M : K(l)]$ is clopen in $\text{Gal}[M : K]$. Therefore P is clopen in $\text{Gal}[L : K]$, because p_M is continuous. Finally, the homeomorphism above implies that $\text{ev}_l^{-1}(\{l_0\})$ is a clopen in $\text{Gal}[L : K]$.

The initial topology of the statement is the coarsest topology such that each mapping ev_l is continuous. The first part of the proof indicates already that the topology of Definition 4.5 is finer than the initial topology of the statement. It thus remains to prove that each fundamental neighborhood $\text{Gal}[L : M]$ of id_L for the topology in Definition 4.5 (see Proposition 4.8) is a neighborhood of id_L for the initial topology of the statement.

Indeed, the finite-dimensional Galois subextension M is generated by finitely many elements $l_1, \ldots, l_n$. For each element $f \in \text{Gal}[L : K]$, the condition $f \in \text{Gal}[L : M]$ is equivalent to $f(l_1) = l_1, \ldots, f(l_n) = l_n$. This is further equivalent to

$$f \in \text{ev}_{l_1}^{-1}(\{l_1\}) \cap \cdots \cap \text{ev}_{l_n}^{-1}(\{l_n\}).$$

Thus
$$\text{Gal}[L : M] = \text{ev}_{l_1}^{-1}(\{l_1\}) \cap \cdots \cap \text{ev}_{l_n}^{-1}(\{l_n\})$$

and this intersection is a neighborhood of id_L for the initial topology of the statement. $\qquad\square$

4.3 The Classical Galois Theorem in Arbitrary Dimension

This section generalizes the classical Galois theorem (see Theorem 1.24) to the case of an arbitrary Galois extension $K \subseteq L$ of fields. We use freely the notation of Section 1.3.

Theorem 4.10 (Galois theorem) *Let $K \subseteq L$ be a Galois extension of fields. The mappings*

$$K \subseteq M \subseteq L \mapsto \mathsf{Gal}[L : M], \quad G \subseteq \mathsf{Gal}[L : K] \mapsto \mathsf{Fix}(G)$$

describe a contravariant isomorphism between

- *the lattice of arbitrary intermediate field extensions $K \subseteq M \subseteq L$;*
- *the lattice of closed subgroups $G \subseteq \mathsf{Gal}[L : K]$ of the topological Galois group (see Definition 4.5).*

For clarity, we split the proof into several lemmas.

Lemma 4.11 *Let $K \subseteq M \subseteq L$ be a finite-dimensional intermediate field extension. The subgroup $\mathsf{Gal}[L : M]$ is clopen in $\mathsf{Gal}[L : K]$.*

Proof The extension M is finitely generated over K, thus by Proposition 4.3, there exists a finite-dimensional Galois subextension $K \subseteq N \subseteq L$ such that $M \subseteq N$. We have

$$\mathsf{id}_L \in \mathsf{Gal}[L : N] \subseteq \mathsf{Gal}[L : M]$$

and by Proposition 4.8, $\mathsf{Gal}[L : N]$ is a clopen neighborhood of id_L.

In a topological group, every subgroup containing an open subgroup is itself open; moreover, every open subgroup is closed. Let us prove this explicitly in the situation of the present lemma.

For each $f \in \mathsf{Gal}[L : M]$, multiplication by f is a homeomorphism which maps id_L to f and each clopen subgroup $\mathsf{Gal}[L : N]$ (see Proposition 4.8) to a clopen subset $U_f \subseteq \mathsf{Gal}[L : M]$ containing f. Thus $\mathsf{Gal}[L : M]$ is open since it contains a neighborhood U_f of each of its points f.

Let us prove that $\mathsf{Gal}[L : M]$ is also closed. Given $g \notin \mathsf{Gal}[L : M]$, it suffices to prove that $U_g \cap \mathsf{Gal}[L : M] = \emptyset$, since we know already that U_g is open and contains g. If not, it would suffice to choose $h = g \circ h'$ in this intersection to get $g = h \circ h'^{-1}$ with $h, h' \in \mathsf{Gal}[L : M]$, thus $g \in \mathsf{Gal}[L : M]$. This would be a contradiction. □

Lemma 4.12 *Let $K \subseteq M \subseteq L$ be an arbitrary subextension. The subgroup $\mathsf{Gal}[L : M]$ is closed in $\mathsf{Gal}[L : K]$.*

Proof We have

$$\begin{aligned}
\mathsf{Gal}[L : M] &= \{f \in \mathsf{Gal}[L : K] \,|\, \forall m \in M \; f(m) = m\} \\
&= \{f \in \mathsf{Gal}[L : K] \,|\, \forall m \in M \; f \in \mathsf{Gal}[L : K(m)]\} \\
&= \bigcap_{m \in M} \mathsf{Gal}[L : K(m)].
\end{aligned}$$

Each subextension $K(m)$ is finite-dimensional over K (see Proposition 1.7). By Lemma 4.11, $\mathsf{Gal}[L : M]$ is closed as an intersection of closed subsets. □

Lemma 4.13 *Let $G \subseteq \mathsf{Gal}[L : K]$ be a closed subgroup. Let $f \in \mathsf{Gal}[L : K]$ be such that in each finite-dimensional Galois subextension $K \subseteq M \subseteq L$, there exists a $g \in G$ such that $f|_M = g|_M$. Then $f \in G$.*

Proof Let $K \subseteq N \subseteq L$ be a finite-dimensional subextension generated by the elements $l_1, \ldots, l_n$; Proposition 4.9 implies that

$$
\begin{aligned}
V_N(f) &= \left\{ g \in \mathsf{Gal}[L : K] \middle| g|_N = f|_N \right\} \\
&= \left\{ g \in \mathsf{Gal}[L : K] \middle| g(l_1) = f(l_1), \cdots, g(l_n) = f(l_n) \right\} \\
&= \mathsf{ev}_{l_1}^{-1} \left(f(l_1) \right) \cap \cdots \cap \mathsf{ev}_{l_n}^{-1} \left(f(l_n) \right)
\end{aligned}
$$

is a neighborhood of f.

These neighborhoods $V_N(f)$ constitute a fundamental system of neighborhoods of f. Indeed, again by Proposition 4.9, each neighborhood of f contains a neighborhood of the form

$$
f \in \mathsf{ev}_{a_1}^{-1}(b_1) \cap \cdots \cap \mathsf{ev}_{a_m}^{-1}(b_m)
$$

where $a_i, b_i \in L$. In particular $f(a_i) = b_i$ and

$$
\begin{aligned}
\mathsf{ev}_{a_1}^{-1}(b_1) &\cap \cdots \cap \mathsf{ev}_{a_m}^{-1}(b_m) \\
&= \left\{ g \in \mathsf{Gal}[L : K] \middle| g(a_1) = b_1, \ldots, g(a_m) = b_m \right\} \\
&= \left\{ g \in \mathsf{Gal}[L : K] \middle| g(a_1) = f(a_1), \ldots, g(a_m) = f(a_m) \right\} \\
&= V_{K(a_1, \ldots, a_n)}(f).
\end{aligned}
$$

By Proposition 4.3, each subextension $K \subseteq N \subseteq L$ generated by finitely many elements is contained in a finite-dimensional Galois subextension $K \subseteq M \subseteq L$. The inclusion $N \subseteq M$ implies

$$
f \in V_M(f) \subseteq V_N(f).
$$

Thus the subsets $V_M(f)$, with M a finite-dimensional Galois extension, still constitute a fundamental system of neighborhoods of f.

Let now $f \in \mathsf{Gal}[L : K]$ be such that for each finite-dimensional Galois subextension $K \subseteq M \subseteq L$, there exists a $g \in G$ such that $f|_M = g|_M$. This means that for each M of this type, $V_M(f) \cap G \neq \emptyset$. Since the $V_M(f)$ constitute a fundamental system of neighborhoods of f, we get $f \in \overline{G}$. And since G is closed, $f \in \overline{G} = G$. $\square$

Lemma 4.14 *Each closed subgroup $G \subseteq \mathsf{Gal}[L : K]$ is such that $G = \mathsf{Gal}\big[L : \mathsf{Fix}(G)\big]$.*

Proof The elements of G are field homomorphisms, thus $\mathsf{Fix}(G)$ is a field. The inclusions

$$
G \subseteq \mathsf{Gal}[L : \mathsf{Fix}(G)] \subseteq \mathsf{Gal}[L : K]
$$

are obvious. By Proposition 1.14, $\mathsf{Fix}(G) \subseteq L$ is a Galois extension.

Consider now the subgroup

$$
H_M = \left\{ f|_M \middle| f \in G \right\} \subseteq \mathsf{Gal}\big[M : \mathsf{Fix}(G)\big]
$$

where $\mathsf{Fix}(G) \subseteq M \subseteq L$ is a finite-dimensional intermediate Galois extension. By assumption

$$\begin{aligned}
\mathrm{Fix}(H_M) &= \{m \in M \mid \forall h \in H_M \;\; h(m) = m\} \\
&= \{m \in M \mid \forall f \in G \;\; f(m) = m\} \\
&= \mathrm{Fix}(G).
\end{aligned}$$

The classical Galois Theorem (see Theorem 1.24) implies

$$H_M = \mathrm{Gal}\big[M : \mathrm{Fix}(H_M)\big] = \mathrm{Gal}\big[M : \mathrm{Fix}(G)\big].$$

In particular, for each element $f \in \mathrm{Gal}\big[L : \mathrm{Fix}(G)\big]$ and each extension M as above, $f|_M$ lies in H_M. Thus for each f and each M as above, there exists a $g \in G$ such that $f|_M = g|_M$. By Lemma 4.13 applied to $\mathrm{Fix}(G)$ instead of K, this forces $f \in G$. $\square$

Lemma 4.15 *For each subextension $K \subseteq M \subseteq L$, $M = \mathrm{Fix}(\mathrm{Gal}[L : M])$.*

Proof It is obvious that

$$M \subseteq \mathrm{Fix}\big(\mathrm{Gal}[L : M]\big) \subseteq L,$$

where $M \subseteq L$ is still a Galois extension (see Proposition 1.14). If $l \in \mathrm{Fix}\big(\mathrm{Gal}[L : M]\big)$, Proposition 4.3 implies the existence of a finite-dimensional Galois subextension

$$M \subseteq M' \subseteq L, \;\; l \in M'.$$

By Proposition 4.2, each M-automorphism of M' is the restriction of an M-automorphism of L. But

$$l \in \mathrm{Fix}\big(\mathrm{Gal}[L : M]\big) \subseteq \mathrm{Fix}\big(\mathrm{Gal}[L : M']\big).$$

Again the classical Galois Theorem 1.24 implies

$$l \in \mathrm{Fix}\big(\mathrm{Gal}[M' : M]\big) = M.$$ $\square$

4.4 Profinite G-Sets

As we have seen (see Theorem 4.10), generalizing the Galois theorem to the case of an arbitrary Galois extension $K \subseteq L$ of fields requires providing the Galois group with the structure of a topological group, more precisely, a profinite group. The Grothendieck approach to the Galois theorem (see Theorem 2.37) refers to sets provided with an action of the Galois group; its arbitrary dimension version will require in the same way the introduction of profinite topologies on the involved $\mathrm{Gal}[L : K]$-sets.

Definition 4.16 Let G be a topological group. A G-space is a topological space X provided with a continuous G-action which makes it a G-set

$$G \times X \longrightarrow X, \quad (g,x) \mapsto gx.$$

A morphism $f \colon X \longrightarrow Y$ of G-spaces is a continuous morphism of G-sets.

In this book, we shall only be interested in the profinite G-spaces on a profinite group G. We write G-Prof for the corresponding category.

Proposition 4.17 *Let G be a profinite group and $H \subseteq G$ a subgroup. The following conditions are equivalent:*

1. *H is closed in G;*
2. *G/H is a profinite G-space for the quotient topology.*

Proof $(1 \Rightarrow 2)$. G is compact, thus G/H with the quotient topology is compact, as a direct image of a compact space. It remains to prove that G/H is totally disconnected.

If $x \in G$, its equivalence class $[x]$ is closed, since it is the image of the closed subset H under the homeomorphism

$$G \longrightarrow G, \quad g \mapsto gx.$$

If $[x] \neq [y]$ are distinct equivalence classes, $[x]$ and $[y]$ are thus disjoint closed subsets. By Proposition 3.15, they are contained in two disjoint clopens $[x] \subseteq U$, $[y] \subseteq V$.

The saturation $\widetilde{U}$ of U for the equivalence relation generated by H

$$\widetilde{U} = \{hz \mid h \in H, \, z \in U\}$$

is the image of $H \times U$ under the multiplication of G. But $H \times U$ is closed, thus compact in $G \times G$; therefore $\widetilde{U}$ is compact in G. The saturation $\widetilde{U}$ is also the union of all the subsets hU, with $h \in H$. Since each mapping

$$G \longrightarrow G, \quad z \mapsto hz$$

is a homeomorphism, each hU is clopen. Since $\widetilde{U}$ is compact, there exists a finite union

$$\widetilde{U} = h_1 U \cup \cdots \cup h_n U, \quad h_1, \ldots, h_n \in H.$$

Thus the saturation $\widetilde{U}$ of U is still clopen.

Observe next that $\widetilde{U}$ is disjoint from $[y]$, because U is disjoint from $[y]$. Indeed, $hu = h'y$ with $h, h' \in H$, $u \in U$ would imply $u = h^{-1}h'y \in [y]$. Thus $\widetilde{U}$ and $\complement \widetilde{U}$ are two open, closed, disjoint and saturated subsets which contain respectively $[x]$ and $[y]$. By saturation, their images in the quotient G/H are thus two disjoint clopen subsets which contain respectively the points $[x]$ and $[y]$. The quotient is thus profinite.

We must still prove that the action of G on G/H is continuous. Consider the following commutative diagram

$$G \times G \xrightarrow{\mu} G$$

$$\text{id} \times p \downarrow \qquad \qquad \downarrow p$$

$$G \times G/H \cdots\cdots\overset{}{\underset{\mu_H}{\cdots}}\cdots\!\!> G/H$$

where μ and μ_H are the G-actions. The action μ_H is continuous because μ and p are continuous and $G \times G/H$ has the quotient topology of $G \times G$ (see Proposition 3.16).

$(2 \Rightarrow 1)$. If G/H with the quotient topology is profinite, it is in particular Hausdorff; thus $\{[0]\}$ is closed in G/H and therefore $p^{-1}([0]) = H$ is closed in G. $\quad \square$

Every discrete finite space X is profinite. But when G is a non-discrete profinite group, being a finite discrete G-space refers now to the continuity of the action

$$G \times X \longrightarrow X$$

where the topology of $G \times X$ is not discrete, since the topology of G is not.

Proposition 4.18 *Let $K \subseteq L$ be a Galois extension of fields. For each finite-dimensional Galois subextension $K \subseteq M \subseteq L$, there exists a functor*

$$\gamma_M : \text{Gal}[M : K]\text{-Set}_f \longrightarrow \text{Gal}[L : K]\text{-Disc}_f, \quad X \longrightarrow X$$

where

- *$\text{Gal}[M : K]$ is the classical Galois group (see Definition 1.11);*
- *$\text{Gal}[L : K]$ is the topological Galois group (see Definition 4.5);*
- *$\text{Gal}[M : K]\text{-Set}_f$ is the category of finite $\text{Gal}[M : K]$-sets;*
- *$\text{Gal}[L : K]\text{-Disc}_f$ is the category of finite discrete $\text{Gal}[L : K]$-spaces;*
- *the action of $g \in \text{Gal}[L : K]$ on $x \in \text{Gal}[M : K]\text{-Set}_f$ is $gx = g|_M x$;*
- *each functor γ_M is full and faithful;*
- *up to equivalences, the category $\text{Gal}[L : K]\text{-Disc}_f$ is the filtered union of its full subcategories $\text{Gal}[M : K]\text{-Set}_f$.*

Proof For ease of notation, we write $G = \text{Gal}[L : K]$ and $G_M = \text{Gal}[M : K]$. We know that $G = \lim_M G_M$, where the limit is cofiltered, each G_M is finite discrete and the projections of the limit are topological quotients (see Proposition 4.6).

If $X \in G_M\text{-Set}_f$, the action of G on $\gamma_M(X)$ is continuous because it is equal to the composite

$$G \times X \xrightarrow{p_M \times \text{id}} G_M \times X \xrightarrow{\mu_M} X$$

where G_M and X are discrete, μ_M is the action of G_M on X and p_M is a projection of the limit.

To prove that γ_M is full and faithful, we consider an arbitrary mapping $f : X \longrightarrow Y$ between two G_M-sets. In the diagram

$$G \times X \xrightarrow{p_M \times \text{id}} G_M \times X \xrightarrow{\mu_M} X$$

$$\text{id} \times f \downarrow \qquad \text{id} \times f \downarrow \qquad\qquad \downarrow f$$

$$G \times Y \xrightarrow{p_M \times \text{id}} G_M \times Y \xrightarrow{\mu_M} Y$$

the left-hand square is commutative. Since p_M is surjective, the commutativity of the external rectangle is equivalent to the commutativity of the right-hand square. This means precisely that f is a morphism of G-sets, thus of discrete G-spaces, if and only if it is a morphism of G_M-sets.

To prove the last assertion, choose $X \in G\text{-Disc}_f$. The canonical morphism

$$\left(\lim_M G_M \right) \times X \xrightarrow{\;\cong\;} \lim_M (G_M \times X)$$

is a homeomorphism because, in every category, products commute trivially with limits. We can also write X as a constant limit $X = \lim_M X$, because the diagram of indices is cofiltered, thus connected. This implies

$$\lim_M (G_M \times X) \cong \left(\lim_M G_M \right) \times \left(\lim_M X \right) \cong \left(\lim_M G_M \right) \times X.$$

By the Stone duality (see Theorem 3.29), the dual of Proposition 3.32 implies

$$\text{Cont} \left(\lim_M (G_M \times X), X \right) \cong \operatorname{colim}_M \text{Cont}(G_M \times X, X)$$

because X is a finite discrete space. Thus the composite

$$\lim_M (G_M \times X) \xrightarrow{\;\cong\;} \left(\lim_M G_M \right) \times X \xrightarrow{\;\cong\;} G \times X \xrightarrow{\mu} X,$$

where μ is the G-space structure of X, admits a factorization

$$\lim_M (G_M \times X) \xrightarrow{p_M \times \text{id}} G_{M_0} \times X \xrightarrow{\mu_{M_0}} X.$$

This means that for each $g \in G$ and $x \in X$, $\mu(g, x) = \mu_{M_0}(g|_{M_0}, x)$. The mapping μ_{M_0} is a G_{M_0}-set structure on X, because μ is a G-set structure and $p_M \times \text{id}$ is surjective. This proves that $(X, \mu) = \gamma_{M_0}(X, \mu_{M_0})$. $\square$

Proposition 4.19 *Let $K \subseteq L$ be a Galois extension of fields. Each profinite* Gal$[L : K]$*-space is a cofiltered limit of finite discrete* Gal$[L : K]$*-spaces.*

Proof Let X be a profinite Gal$[L : K]$-space. We can write X as the cofiltered limit of its finite discrete quotients (see Theorem 3.18): $X = \lim_{i \in I} X_i$. In general, an action on X has no reason to induce an action on X_i. It is straightforward that the limit $X = \lim_{i \in I} X_i$ is isomorphic to the limit $X = \lim_{j \in J} X_j$ for each subset $J \subseteq I$ with the property

$$\forall i \in I \ \exists j \in J \ \ j \le i.$$

It thus suffices to verify that the set J of indices $j \in I$ such that X induces a Gal$[L : K]$-set structure on X_j has this property.

Let R_j be the equivalence relation on X such that $X_j = X/R_j$. When

$$(x, y) \in R_j \Rightarrow \forall g \in G \ (gx, gy) \in R_j,$$

the action of Gal$[L : K]$ on X induces an action on the quotient. This action is continuous because the following diagram is commutative

$$
\begin{array}{ccc}
G \times X & \xrightarrow{\ \mu\ } & X \\[2pt]
{\scriptstyle \text{id} \times p_j}\Big\downarrow & & \Big\downarrow{\scriptstyle p_j} \\[2pt]
G \times X_j & \cdots\!\!\xrightarrow{\ \mu_j\ }\!\!\cdots\!\!> & X_j
\end{array}
$$

where μ and μ_j are the actions. The action μ_j is thus continuous because μ and p_j are continuous and $G \times X_j$ has the quotient topology of $G \times X$ (see Proposition 3.16).

Given an arbitrary index $i \in I$, let us consider the following relation R_j on X

$$(x, y) \in R_j \text{ if and only if } \forall g \in G \ (gx, gy) \in R_i.$$

This relation R_j is trivially an equivalence relation. Putting $g = 1$, we observe that $R_j \subseteq R_i$. And trivially

$$(x, y) \in R_j \Rightarrow \forall g \in G \ (gx, gy) \in R_j.$$

To conclude, we have to check that X/R_j is still a finite discrete space.

Let us consider the two continuous functions

$$\alpha, \beta \colon G \times X \times X \overset{\longrightarrow}{\underset{\longrightarrow}{\ }} X \times X$$

where

$$\alpha(g, x, y) = (x, y), \quad \beta(g, x, y) = (gx, gy).$$

We observe that

$$(x, y) \notin R_j \Leftrightarrow \exists g \in G \ (gx, gy) \notin R_i,$$

that is,

$$(X \times X) \setminus R_j = \alpha\Big(\beta^{-1}\big((X \times X) \setminus R_i\big)\Big).$$

The equivalence relation R_i is open in $X \times X$. Indeed we have the following pullback

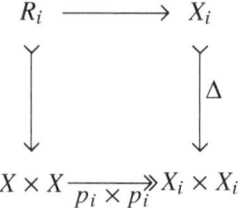

where the diagonal Δ is open, because X_i is discrete. This implies that $\beta^{-1}((X \times X) \setminus R_i)$ is closed, thus compact in $G \times X \times X$. So $(X \times X) \setminus R_j$ is compact, thus closed in $X \times X$. This proves that the equivalence relation R_j is open in $X \times X$.

Let us now consider the pullback

$$
\begin{array}{ccc}
[x] & \longrightarrow & R_j \\
\downarrow & & \downarrow \\
X & \xrightarrow[(\mathrm{id},\, \ulcorner x \urcorner)]{} & X \times X
\end{array}
$$

where $\ulcorner x \urcorner$ is the constant mapping on the fixed element $x \in X$. This proves that each equivalence class $[x]$ is open in X. Therefore the quotient X/R_j, with the quotient topology, is a discrete space (each point is open). But this quotient is also compact, as a continuous image of the compact space X. It is thus finite.   □

4.5 Arbitrary Split Algebras

Again, we study the arbitrary split algebras (see Definition 2.23) in terms of their finite-dimensional subalgebras split by a finite-dimensional extension of fields.

Proposition 4.20 *Let K be a field. Every algebraic K-algebra A is the filtered union of its finite-dimensional subalgebras.*

Proof The finite-dimensional K-algebras are exactly the finitely generated K-subalgebras (see Proposition 2.22). The finitely generated K-subalgebras obviously constitute a filtered poset, whose set-theoretical union is the whole algebra, because each element $a \in A$ belongs to the finitely generated subalgebra $K(a)$. This set-theoretical union is also trivially a union in the category of K-algebras.   □

Let us recall that every subalgebra of a split algebra is itself split (see Proposition 2.26).

Proposition 4.21 *Let $K \subseteq L$ be a Galois extension of fields and A a K-algebra split by that extension.*

1. *For each finite-dimensional K-subalgebra $B \subseteq A$, there exists a finite-dimensional Galois subextension $K \subseteq M \subseteq L$ such that B is split by $K \subseteq M$.*
2. *For each such subextension, composition with the inclusion $M \subseteq L$ induces a bijection*

$$\mathsf{Hom}_K(B, M) \cong \mathsf{Hom}_K(B, L)$$

between finite sets.

Proof The K-subalgebra B is generated over K by finitely many elements $b_1, \ldots, b_n$. Each element b_i admits a minimal polynomial $p_i(X) \in K[X]$, and $p_i(X)$ has in L the roots $l_1^i, \ldots, l_{m_i}^i$. By Proposition 4.1, the extension

$$M_i = K(l_1^i, \ldots, l_{m_i}^i)$$

is a finite-dimensional Galois extension. Proposition 4.3 implies the existence of a finite-dimensional Galois subextension $K \subseteq M \subseteq L$ containing $M_1, \ldots, M_n$. By Theorem 2.27, each K-algebra $K(b_i)$ is split by the extension $K \subseteq M$ because

$$\#\mathsf{Hom}_K\big(K(b_i), M\big) = \#\mathsf{Hom}_K\left(\frac{K[X]}{\langle p_i(X)\rangle}, M\right)$$
$$= \text{number of roots of } p_i(X) \text{ in } M$$
$$= \text{degree of } p_i(X)$$
$$= \dim {}_K K(b_i),$$

by Proposition 2.12. Lemma 2.29 implies that $K(b_1) \cup \cdots \cup K(b_n) = B$ is split by $K \subseteq M$.

Now each K-homomorphism $f : B \longrightarrow L$ is such that, for each element $b \in B$ with minimal polynomial $p(X) \in K[X]$,

$$p\big(f(b)\big) = f\big(p(b)\big) = f(0) = 0.$$

Thus $f(b)$ is a root of $p(X)$ in L and since $K \subseteq M$ is a Galois extension, $f(b) \in M$. This proves the isomorphism $\mathsf{Hom}_K(B, L) \cong \mathsf{Hom}_K(B, M)$. Theorem 2.27 asserts that these sets are finite.   □

Proposition 4.22 *Let $K \subseteq L$ be a Galois extension of fields. For each K-algebra A split by this extension, there exists a bijection*

$$\mathsf{Hom}_K(A, L) \cong \lim_{B} \mathsf{Hom}_K(B, L)$$

where

- *the indices B are the finite-dimensional subalgebras $B \subseteq M$;*
- *the projection p_B of the limit is the restriction to $B \subseteq A$;*

- *each term $\mathrm{Hom}_K(B, L)$ of the limit is a finite set;*
- *the limit is cofiltered.*

Proof By Lemma 4.3, A is the filtered union – thus the colimit – of its finite-dimensional K-subalgebras; this implies

$$\mathrm{Hom}_K(A, L) \cong \mathrm{Hom}_K\left(\underset{B}{\mathrm{colim}}\, B, L\right) \cong \underset{B}{\lim}\, \mathrm{Hom}_K(B, L)$$

and the limit is cofiltered. The terms of the limit are finite by Proposition 4.21. □

Proposition 4.23 *Let K be a field and $A = \bigcup_B B$ an algebraic K-algebra presented as the filtered union of all its finite-dimensional K-subalgebras (see Proposition 4.20). For each finite-dimensional K-algebra C, the canonical morphism*

$$\rho\colon \underset{B}{\mathrm{colim}}\, \mathrm{Hom}_K(C, B) \overset{\cong}{\longrightarrow} \mathrm{Hom}_K(C, A)$$

is bijective.

Proof The canonical morphism ρ of the statement is induced by the various morphisms $\mathrm{Hom}_K(C, B) \subseteq \mathrm{Hom}_K(C, A)$ of composition with the inclusions $B \subseteq A$.

Let $c_1, \ldots, c_n$ be a base of C as a K-vector space. A morphism of K-algebras $f\colon C \longrightarrow A \cong \mathrm{colim}_B B$ is such that

$$\forall i = 1, \ldots, n \quad f(c_i) = b_i \quad \text{with} \quad b_i \in B_i.$$

By Proposition 4.20, we can choose $B_0 \subseteq A$ finite-dimensional and containing all the elements b_i. This implies the existence of a K-linear factorization g

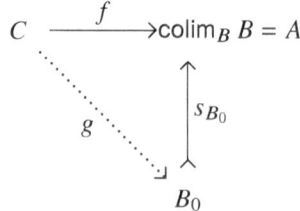

where

$$g\left(\sum_{i=1,\ldots,n} k_i c_i\right) = \sum_{i=1,\ldots,n} k_i b_i.$$

Since s_{B_0} is injective and f is a homomorphism of K-algebras, g is a homomorphism of K-algebras. The diagram above shows that $f = \rho([g])$. This proves that ρ is surjective.

To prove that ρ is injective, consider $g'\colon C \longrightarrow B_1$, another homomorphism such that $f = \rho([g'])$. Again by Proposition 4.20, we can consider g and g' as taking values in the same finite-dimensional K-subalgebra $B_2 \subseteq A$. This implies $g = g'$ as mappings with values in B_2, because s_{B_2} is injective. Finally, this implies $[g] = [g']$ in the colimit. □

4.6 The More General Grothendieck Galois theorem

We are now ready to extend the Grothendieck Galois Theorem 2.37 to the case of an arbitrary Galois extension $K \subseteq L$ of fields.

Theorem 4.24 (Galois Theorem) *Let $K \subseteq L$ be a Galois extension of fields. There exists a contravariant equivalence of categories*

$$\mathsf{Hom}_K(-, L)\colon \mathsf{Split}[L : K] \longrightarrow \mathsf{Gal}[L : K]\text{-}\mathsf{Prof}$$

where

- $\mathsf{Split}[L : K]$ *is the category of K-algebras split by the extension $K \subseteq L$;*
- $\mathsf{Gal}[L : K]\text{-}\mathsf{Prof}$ *is the category of profinite $\mathsf{Gal}[L : K]$-spaces.*

Again, for clarity, we split the proof into various lemmas.

Lemma 4.25 *Let $A \in \mathsf{Split}[L : K]$ and $B \subseteq A$ a finite-dimensional K-subalgebra. The mapping*

$$\mathsf{Gal}[L : K] \times \mathsf{Hom}_K(B, L) \longrightarrow \mathsf{Hom}_K(B, L), \quad (g, f) \mapsto g \circ f$$

is a structure of $\mathsf{Gal}[L : K]$-space on the discrete finite set $\mathsf{Hom}_K(B, L)$.

Proof By Proposition 4.21, there exists a finite-dimensional Galois subextension $K \subseteq M \subseteq L$ such that B is split by $K \subseteq M$. The following is a $\mathsf{Gal}[M : K]$-set structure on $\mathsf{Hom}_K(B, M)$

$$\mathsf{Gal}[M : K] \times \mathsf{Hom}_K(B, M) \longrightarrow \mathsf{Hom}_K(B, M), \quad (g, f) \mapsto g \circ f.$$

Proposition 4.18 implies that

$$\mathsf{Gal}[L : K] \times \mathsf{Hom}_K(B, M) \longrightarrow \mathsf{Hom}_K(B, M), \quad (g, f) \mapsto g|_M \circ f$$

is a discrete $\mathsf{Gal}[L : K]$-space structure. But by Proposition 4.21, $\mathsf{Hom}_K(B, L) \cong \mathsf{Hom}_K(B, M)$. $\qquad\square$

Lemma 4.26 *If $A \in \mathsf{Split}[L : K]$, the mapping*

$$\mu\colon \mathsf{Gal}[L : K] \times \mathsf{Hom}_K(A, L) \longrightarrow \mathsf{Hom}_K(A, L), \quad (g, f) \mapsto g \circ f$$

is a profinite $\mathsf{Gal}[L : K]$-space structure on $\mathsf{Hom}_K(A, L)$.

Proof The statement describes exactly the limit, cofiltered by Proposition 4.21, of the situations in Lemma 4.25. $\qquad\square$

Lemma 4.27 *The construction above extends as a functor*

$$\mathsf{Hom}_K(-, L)\colon \mathsf{Split}[L : K] \longrightarrow \mathsf{Gal}[L : K]\text{-}\mathsf{Prof}.$$

Proof Let $\alpha \colon A \longrightarrow A'$ be a morphism in $\mathsf{Split}[L : K]$. We must prove that the mapping

$$\mathsf{Hom}_K(\alpha, L) \colon \mathsf{Hom}_K(A', L) \longrightarrow \mathsf{Hom}_K(A, L), \quad h \mapsto h \circ \alpha$$

is a continuous morphism of $\mathsf{Gal}[L : K]$-spaces. Being a morphism of $\mathsf{Gal}[L : K]$-sets reduces to the associativity of the composition (see Lemma 4.26)

To prove the continuity, by definition of a limit, it suffices to show that for each finite-dimensional K-subalgebra $B \subseteq A$, the composite

$$\mathsf{Hom}_K(A', L) \xrightarrow{\mathsf{Hom}_K(\alpha, L)} \mathsf{Hom}_K(A, L) = \lim_B \mathsf{Hom}_K(B, L) \xrightarrow{p_B^A} \mathsf{Hom}_K(B, L)$$

is continuous. Since B is finite-dimensional, $\alpha(B) \subseteq A'$ is a finite-dimensional K-subalgebra. The following diagram is commutative

$$
\begin{array}{ccc}
\mathsf{Hom}_K(A', L) & \xrightarrow{\quad \mathsf{Hom}_K(\alpha, L) \quad} & \mathsf{Hom}_K(A, L) \\
{\scriptstyle p_{\alpha(B)}^{A'}} \downarrow & & \downarrow {\scriptstyle p_B^A} \\
\mathsf{Hom}_K(\alpha(B), L) & \xrightarrow[\mathsf{Hom}_K(\alpha|_B, L)]{} & \mathsf{Hom}_K(B, L)
\end{array}
$$

where the vertical morphisms are continuous by definition and $\mathsf{Hom}_K(\alpha|_B, L)$ is continuous because it is defined between discrete spaces. $\square$

Lemma 4.28 *The functor $\mathsf{Hom}_K(-, L)$ is full and faithful.*

Proof Consider $A, A' \in \mathsf{Split}[L : K]$. By Definition 2.23 and Proposition 4.20, we can write

$$A = \bigcup_B B \equiv \underset{B}{\mathrm{colim}}\, B, \quad A' = \bigcup_{B'} B' \equiv \underset{B'}{\mathrm{colim}}\, B',$$

where $B \subseteq A$, $B' \subseteq A'$ are finite-dimensional K-subalgebras and the colimits are filtered. For each pair B, B', by Propositions 4.21 and 4.3, we can choose a finite-dimensional Galois subextension $K \subseteq M_{B,B'} \subseteq L$ such that B and B' are split by $K \subseteq M_{B,B'}$. By Proposition 4.21,

$$\mathsf{Hom}_K(B, L) \cong \mathsf{Hom}_K(B, M_{B,B'}), \quad \mathsf{Hom}_K(B', L) \cong \mathsf{Hom}_K(B', M_{B,B'}).$$

Observe now that

$$\mathsf{Hom}\big(\lim_B \mathsf{Hom}_K(B, L), \mathsf{Hom}_K(B', L)\big)$$

$$\cong \underset{B}{\mathrm{colim}}\, \mathsf{Hom}\big(\mathsf{Hom}_K(B, L), \mathsf{Hom}_K(B', L)\big).$$

By the Stone duality 3.29, the dual of Proposition 3.32 implies

$$\mathsf{Cont}\Big(\varprojlim_B \mathsf{Hom}_K(B,L), \mathsf{Hom}_K(B',L)\Big)$$

$$\cong \mathop{\mathrm{colim}}_M \mathsf{Cont}\Big(\mathsf{Hom}_K(B,L), \mathsf{Hom}_K(B',L)\Big)$$

because $\mathsf{Hom}_K(B',L)$ is a finite discrete space. It is obvious that this bijection transforms a morphism of $\mathsf{Gal}[L:K]$-sets into the equivalence class of a morphism of $\mathsf{Gal}[L:K]$-sets, and conversely. This proves the announced bijection.

We obtain now

$$\mathsf{Hom}\big(\mathsf{Hom}_K(A,L), \mathsf{Hom}_K(A',L)\big)$$
$$\cong \mathsf{Hom}\big(\mathsf{Hom}_K(\mathrm{colim}_B\, B, L), \mathsf{Hom}_K(\mathrm{colim}_{B'}\, B', L)\big)$$
$$\cong \mathsf{Hom}\big(\varprojlim_B \mathsf{Hom}_K(B,L), \varprojlim_{B'} \mathsf{Hom}_K(B',L)\big)$$
$$\cong \varprojlim_{B'} \mathsf{Hom}\big(\varprojlim_B \mathsf{Hom}_K(B,L), \mathsf{Hom}_K(B',L)\big)$$
$$\cong \varprojlim_{B'} \mathrm{colim}_B\, \mathsf{Hom}\big(\mathsf{Hom}_K(B,L), \mathsf{Hom}_K(B',L)\big)$$
$$\cong \varprojlim_{B'} \mathrm{colim}_B\, \mathsf{Hom}\big(\mathsf{Hom}_K(B,M_{BB'}), \mathsf{Hom}_K(B',M_{BB'})\big)$$
$$\cong \varprojlim_{B'} \mathrm{colim}_B\, \mathsf{Hom}(B',B) \quad \text{by Theorem 2.37}$$
$$\cong \varprojlim_{B'} \mathsf{Hom}(B', \mathrm{colim}\, B) \quad \text{by Proposition 4.23}$$
$$\cong \mathsf{Hom}(\mathrm{colim}\, B', \mathrm{colim}\, B)$$
$$\cong \mathsf{Hom}(A', A).$$

This proves that $\mathsf{Hom}_K(-,L)$ is full and faithful.   □

Lemma 4.29 *The functor* $\mathsf{Hom}_K(-,L)$ *is essentially surjective on the objects.*

Proof Each profinite $\mathsf{Gal}[L:K]$-space X is a cofiltered limit $X \cong \varprojlim_{i\in I} X_i$ of finite discrete $\mathsf{Gal}[L:K]$-spaces X_i (see Proposition 4.19). By Proposition 4.18, each X_i is a $\mathsf{Gal}[M_i:K]$-set for some finite-dimensional Galois subextension $K \subseteq M_i \subseteq L$. By Theorem 2.37, $X_i = \mathsf{Hom}_K(B_i, M_i)$ for some finite-dimensional K-algebra B_i split by $K \subseteq M_i$. By Proposition 4.21

$$X_i \cong \mathsf{Hom}_K(B_i, M_i) \cong \mathsf{Hom}_K(B_i, L).$$

If $f_{ij}: X_i \longrightarrow X_j$ is a morphism in the diagram defining X, we can choose a finite-dimensional Galois subextension $K \subseteq M_{i,j} \subseteq L$ which contains M_i and M_j (see Proposition 4.3). The K-algebras B_i and B_j are *a fortiori* split by $K \subseteq M_{i,j}$, thus Proposition 4.21 implies again

$$X_i \cong \mathsf{Hom}_K(B_i, L) \cong \mathsf{Hom}_K(B_i, M_{i,j}), \quad X_j \cong \mathsf{Hom}_K(B_j, L) \cong \mathsf{Hom}_K(B_j, M_{i,j}).$$

In particular, the composition induces $\mathsf{Gal}[M_{i,j}:K]$-set structures on X_i, X_j, and $f_{i,j}: X_i \longrightarrow X_j$ becomes a morphism of $\mathsf{Gal}[M_{i,j}:K]$-sets. Again Theorem 2.37 implies that this morphism

$$f_{ij}: \mathsf{Hom}_K(B_i, M) = X_i \longrightarrow X_j = \mathsf{Hom}_K(B_j, M)$$

is induced by a morphism $h_{i,j}\colon B_j \longrightarrow B_i$ of K-algebras.

Observe that, starting with a cofiltered diagram $(X_i)_{i \in I}$, we constructed a corresponding filtered diagram $(B_i)_{i \in I}$ in the category of K-algebras split by $K \subseteq L$. Let us consider the filtered colimit $A = \mathrm{colim}_{i \in I}\, B_i$ in the category of K-algebras. This colimit is constructed as in the category of sets, because it is filtered. We want to prove that this K-algebra A is split by $K \subseteq L$.

Each element $a \in A$ has the form $[a_i]$ for some index $i \in I$ and some element $a_i \in B_i$. The minimal polynomial $p(X)$ of a_i admits a decomposition into distinct linear factors over L, because B_i is split by $K \subseteq L$ (see Definition 2.23). But $p(a_i) = 0$ in B_i implies $p(a) = 0$ in A; thus the minimal polynomial $q(X) \in K[X]$ of $a \in A$ is a factor of $p(X)$ (see Proposition 2.16). In particular, $q(X)$ admits a decomposition into distinct linear factors over L. This proves that A is split by $K \subseteq L$.

Finally we have

$$\mathrm{Hom}_K(A, L) \cong \mathrm{Hom}_K\left(\mathrm{colim}_{i \in I} B_i, L\right) \cong \lim_{i \in I} \mathrm{Hom}_K(B_i, L) \cong \lim_{i \in I} X_i \cong X.$$

This concludes the proof of the Lemma and also, by Lemma 2.36, that of Theorem 4.24.   □

Corollary 4.30 *Let $K \subseteq L$ be a Galois extension of fields and $A \subseteq L$ a K-subalgebra. The following conditions are equivalent:*

1. *A is a field;*
2. *$A \rightarrowtail L$ is an extremal monomorphism of K-algebras.*[2]

Proof If A is a field, consider the inclusions of K-algebras $A \subseteq B \subseteq L$ with $A \subseteq B$ an epimorphism. We must prove that $A = B$. Of course B is an A-algebra, since $A \subseteq B$. Moreover, $A \subseteq L$ is a Galois extension of fields (see Proposition 1.14). Two A-homomorphisms $f, g\colon B \underset{\longrightarrow}{\longrightarrow} L$ necessarily coincide on A; but they are also K-homomorphisms and since $A \subseteq B$ is an epimorphism of K-algebras, $f = g$. Thus $\mathrm{Hom}_A(B, L)$ is a singleton. By Theorem 4.24, $A = B$.

Conversely, let $A \subseteq \overline{A} \subseteq L$ be the subfield generated by A, that is,

$$\overline{A} = \left\{ \frac{a}{b} \,\middle|\, a \in A,\ 0 \neq b \in A \right\}.$$

Two morphisms $f, g\colon \overline{A} \underset{\longrightarrow}{\longrightarrow} C$ of K-algebras which coincide on A are equal, thus $A \subseteq \overline{A}$ is an epimorphism of K-algebras. This implies $A = \overline{A}$ because $A \subseteq L$ is an extremal monomorphism.   □

---

[2] A monomorphism f is a left cancelable arrow: $fu = fv$ implies $u = v$. The dual notion is that of an epimorphism. A monomorphism f is *extremal* when $f = g \circ h$, with h an epimorphism, implies that h is an isomorphism. For example in the category of topological spaces, the extremal monomorphisms are – up to a homeomorphism – the inclusions of the subspaces provided with the induced topology.

Let us observe that in the case of a finite-dimensional Galois extension $K \subseteq L$, Theorem 4.24 provides a more elaborate equivalence of categories than that of Theorem 2.37, since it also considers the infinite-dimensional split algebras and the infinite profinite spaces. But of course, the equivalence in Theorem 4.24 also contains as a restriction the equivalence of Theorem 2.37.

Finally observe that, as at the end of Chapter 2, it is possible to present the Galois theorem 4.10 as a corollary of Theorem 4.24. This time the argument is:

- the subextensions $K \subseteq M \subseteq L$ correspond to the extremal monomorphisms $M \rightarrowtail L$ in $\mathsf{Split}[L : K]$ (see Corollary 4.30);
- the closed subgroups $H \subseteq \mathsf{Gal}[L : K]$ correspond to the extremal epimorphisms – that is, the quotients – $\mathsf{Gal}[L : K] \twoheadrightarrow \mathsf{Gal}[L : K]/H$ in $\mathsf{Gal}[L : K]$-Prof (see Proposition 4.17).

Part II
The Galois Theory of Rings

Chapter 5
Adjunctions and Monads

Abstract This chapter presents two essential and closely related notions: adjunctions and monads; each adjunction generates a monad, each monad generates an adjunction. Every module on a ring R is an additive group; but conversely, is there a best R-module associated with an additive group? This is the idea of a pair of adjoint functors: two interdependent constructions, between two categories, in both directions. And what is a real vector space? An elegant way to grasp the spirit of that notion is to say: this is a set in which real linear combinations make sense. Like in a monoid, this approach presents an associative composition: a linear combination of linear combinations yields a linear combination. This provides an example of a *monad* on the category of sets: given a set A, one considers the set $T(A)$ of all formal real linear combinations of elements of A; a vector space is a set A provided with an action of $T(A)$, a set in which every formal linear combination has been given a value. Of course, a monad can be defined over every category, not just Set. The Beck criterion characterizes those categories which are categories of algebras for a monad.

5.1 Adjoint Functors

We focus here on the notion of adjoint functors and limit our study to their properties which are useful in this book.

Definition 5.1 Consider two functors

$$L \colon \mathcal{A} \longrightarrow \mathcal{B}, \quad R \colon \mathcal{B} \longrightarrow \mathcal{A}.$$

These constitute a pair of adjoint functors, with L the left adjoint and R the right adjoint, when there exist natural isomorphisms

$$\theta_{A,B} \colon \mathcal{B}\big(L(A), B\big) \cong \mathcal{A}\big(A, R(B)\big)$$

F. Borceux, *Galois Theories of Fields and Rings*, Coimbra Mathematical Texts 2,
https://doi.org/10.1007/978-3-031-58460-2_6

for all objects $A \in \mathcal{A}$ and $B \in \mathcal{B}$. We shall use the notation $L \dashv R$ to indicate such a situation of adjunction.

Example 5.2 Let R be a commutative ring with unit. The functor on the category of R-modules

$$- \otimes_R B \colon \mathsf{Mod}_R \longrightarrow \mathsf{Mod}_R, \quad A \mapsto A \otimes_R B$$

is left adjoint to the functor

$$\mathsf{Lin}(B, -) \colon \mathsf{Mod}_R \longrightarrow \mathsf{Mod}_R, \quad C \mapsto \mathsf{Lin}(B, C)$$

where $\mathsf{Lin}(B, C)$ is the R-module of R-linear mappings.

Proof The bijection

$$\mathsf{Lin}\Big(A \otimes_R B, C\Big) \cong \mathsf{Lin}\big(A, \mathsf{Lin}(B, C)\big)$$

maps $f \colon A \otimes_R B \longrightarrow C$ to

$$A \longrightarrow \mathsf{Lin}(B, C), \quad a \mapsto \big(B \to C, \ b \mapsto f(a \otimes b)\big)$$

and $g \colon A \longrightarrow \mathsf{Lin}(B, C)$ to

$$A \otimes_R B \longrightarrow C, \quad \sum_{i=1}^n na_i \otimes b_i \mapsto \sum_{i=1}^n g(a_i)(b_i). \qquad \square$$

Example 5.3 When $F \colon \mathcal{A} \longrightarrow \mathcal{B}$ is an equivalence of categories (see Definition 2.34), the "inverse equivalence" $G \colon \mathcal{B} \longrightarrow \mathcal{A}$ is both left and right adjoint to F.

Proof Using the natural isomorphisms of Definition 2.34 and Lemma 2.36, one gets at once natural bijections

$$\mathcal{B}\big(B, F(A)\big) \cong \mathcal{A}\big(FG(B), F(A)\big) \cong \mathcal{A}\big(G(B), A\big).$$

Thus G is left adjoint to F. The other assertion is proved in the same way. $\qquad \square$

The following equivalent characterization is often useful:

Proposition 5.4 *Let*

$$L \colon \mathcal{A} \longrightarrow \mathcal{B}, \quad R \colon \mathcal{B} \longrightarrow \mathcal{A}$$

be functors. The following conditions are equivalent:

1. *L and R constitute a pair of adjoint functors $L \dashv R$;*
2. *there exist natural transformations*

$$\varepsilon \colon \mathsf{id}_{\mathcal{A}} \Rightarrow R \circ L, \quad \eta \colon L \circ R \Rightarrow \mathsf{id}_{\mathcal{B}},$$

such that for all objects $A \in \mathcal{A}$ and $B \in \mathcal{B}$, the following diagrams are commutative:

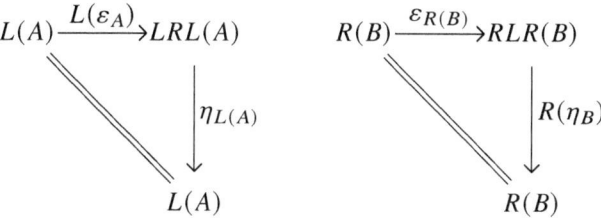

The natural transformation ε is called the unit *of the adjunction and the natural transformation η, the* co-unit *of the adjunction.*

Proof $(1 \Rightarrow 2)$. With the notation of Definition 5.1, it suffices to define

$$\varepsilon_A = \theta_{A,L(A)}\big(\mathrm{id}_{L(A)}\big), \quad \eta_B = \theta^{-1}_{R(B),B}\big(\mathrm{id}_{R(B)}\big).$$

The naturality of θ (and thus θ^{-1}) implies

$$\begin{aligned}
\eta_{L(A)} \circ L(\varepsilon_A) &= \Big(\mathcal{B}\big(L(\varepsilon_A), \mathrm{id}_{L(A)}\big) \circ \theta^{-1}_{RL(A),L(A)}\Big)\big(\mathrm{id}_{RL(A)}\big) \\
&= \Big(\theta^{-1}_{A,L(A)} \circ \mathcal{A}\big(\varepsilon_A, \mathrm{id}_{LR(A)}\big)\Big)\big(\mathrm{id}_{RL(A)}\big) \\
&= \theta^{-1}_{A,L(A)}(\varepsilon_A) \\
&= \mathrm{id}_{L(A)},
\end{aligned}$$

$$\begin{aligned}
R(\eta_B) \circ \varepsilon_{R(B)} &= \Big(\mathcal{A}\big(\mathrm{id}_{R(B)}, R(\eta_B)\big) \circ \theta_{R(B),LR(B)}\Big)\big(\mathrm{id}_{LR(B)}\big) \\
&= \Big(\theta_{R(B),B} \circ \mathcal{B}\big(\mathrm{id}_{LR(B)}, \eta_B\big)\Big)\big(\mathrm{id}_{LR(B)}\big) \\
&= \theta_{R(B),B}(\eta_B) \\
&= \mathrm{id}_{R(B)}.
\end{aligned}$$

The naturality of ε and η is trivial.

$(2 \Rightarrow (1)$. It suffices to define, for each $g \in \mathcal{B}\big(L(A), B\big)$ and $f \in \mathcal{A}\big(A, R(B)\big)$:

$$\theta_{A,B}(g) = R(g) \circ \varepsilon_A, \quad \theta^{-1}_{A,B}(f) = \eta_B \circ L(f).$$

We get at once

$$\big(\theta^{-1}_{A,B} \circ \theta_{A,B}\big)(g) = \eta_B \circ LR(g) \circ L(\varepsilon_A) = g \circ \eta_{L(A)} \circ L(\varepsilon_A) = g$$

$$\big(\theta_{A,B} \circ \theta^{-1}_{A,B}\big)(f) = R(\eta_B) \circ RL(f) \circ \varepsilon_A = R(\eta_B) \circ \varepsilon_{R(B)} \circ f = f.$$

The naturality of θ is trivial. $\qquad\square$

The following result is particularly useful to construct equivalences of categories. We keep using the notation above.

Proposition 5.5 *Let L ⊣ R be adjoint functors*

$$L: \mathcal{A} \longrightarrow \mathcal{B}, \quad R: \mathcal{B} \longrightarrow \mathcal{A}.$$

The following conditions are equivalent:

1. *the functor L is full and faithful;*
2. *the natural transformation $\varepsilon: \mathrm{id}_{\mathcal{A}} \Rightarrow R \circ L$ is an isomorphism.*

And dually:

1. *the functor R is full and faithful;*
2. *the natural transformation $\eta: L \circ R \Rightarrow \mathrm{id}_{\mathcal{B}}$ is an isomorphism.*

Proof If $L \dashv R$, considering the dual categories we get $R^{\mathrm{op}} \dashv L^{\mathrm{op}}$:

$$L^{\mathrm{op}}: \mathcal{A}^{\mathrm{op}} \longrightarrow \mathcal{B}^{\mathrm{op}}, \quad R^{\mathrm{op}}: \mathcal{B}^{\mathrm{op}} \longrightarrow \mathcal{A}^{\mathrm{op}}.$$

The unit and co-unit of the adjunction $L \dashv R$ become the co-unit and the unit of the adjunction $R^{\mathrm{op}} \dashv L^{\mathrm{op}}$. Thus the two statements are dual to each other. Let us prove the second one.

When R is full and faithful, the morphism $\varepsilon_{R(B)}: R(B) \longrightarrow RLR(B)$ has the form $R(\gamma_B)$ for some morphism $\gamma_B: B \longrightarrow LR(B)$. The equality $R(\eta_B) \circ R(\gamma_B) = R(\eta_B) \circ \varepsilon_{R(B)} = \mathrm{id}_{R(B)}$ (see Proposition 5.4) and the faithfulness of R imply $\eta_B \circ \gamma_B = \mathrm{id}_B$. To prove that $\gamma_B \circ \eta_B = \mathrm{id}_{LR(B)}$, observe that

$$\theta_{R(B),LR(B)}(\gamma_B \circ \eta_B) = R(\gamma_B \circ \eta_B) \circ \varepsilon_{R(B)}$$
$$= \varepsilon_{R(B)} \circ R(\eta_B) \circ \varepsilon_{R(B)}$$
$$= \varepsilon_{R(B)}$$
$$= \theta_{R(B),LR(B)}(\mathrm{id}_{LR(B)}).$$

This implies $\gamma_B \circ \eta_B = \mathrm{id}_{LR(B)}$, because $\theta_{R(B),LR(B)}$ is bijective.

Conversely if η is an isomorphism, consider the composite

$$\mathcal{B}(B, B') \xrightarrow{\mathcal{B}(\eta_B, \mathrm{id}_{B'})} \mathcal{B}(LR(B), B') \xrightarrow{\theta_{R(B),B'}} \mathcal{A}(R(B), R(B')).$$

We know that both morphisms are bijective. The image of an element $g \in \mathcal{B}(B, B')$ under this composite is

$$\theta_{R(B),B'}(g \circ \eta_B) = R(g \circ \eta_B) \circ \varepsilon_{R(B)} = R(g) \circ R(\eta_B) \circ \varepsilon_{R(B)} = R(g).$$

This bijective composite is thus the action of R, which is therefore full and faithful.□

Let us conclude this section with an important property of adjoint functors.

Proposition 5.6 *Let* $L \dashv R$ *be adjoint functors*

$$L: \mathcal{A} \longrightarrow \mathcal{B}, \quad R: \mathcal{B} \longrightarrow \mathcal{A}.$$

The functor R preserves all existing limits and the functor L preserves all existing colimits.

Proof Again if $L \dashv R$, considering the dual categories we get $R^{\mathrm{op}} \dashv L^{\mathrm{op}}$:

$$L^{\mathrm{op}}: \mathcal{A}^{\mathrm{op}} \longrightarrow \mathcal{B}^{\mathrm{op}}, \quad R^{\mathrm{op}}: \mathcal{B}^{\mathrm{op}} \longrightarrow \mathcal{A}^{\mathrm{op}}.$$

Thus the two statements are dual to each other. Let us prove the first one.

Let $F: \mathcal{I} \longrightarrow \mathcal{B}$ be a functor admitting the limit

$$\left(p_I: L \longrightarrow F(I)\right)_{I \in \mathcal{I}}$$

in $\mathcal{B}$. If

$$\left(q_I: R(B) \longrightarrow RF(I)\right)_{I \in \mathcal{I}}$$

is a compatible family of morphisms in $\mathcal{A}$, via the natural bijections of the adjunction (see Definition 5.1) we obtain a compatible family in $\mathcal{B}$

$$\left(q_I': LR(B) \longrightarrow F(I)\right)_{I \in \mathcal{I}}$$

and thus a factorization $f: LR(B) \longrightarrow L$ through the limit of F. Again the natural bijections of the adjunction imply that the unique factorization f in $\mathcal{B}$ corresponds to a unique factorization $g: R(B) \longrightarrow R(L)$ in $\mathcal{A}$ of the morphisms q_I through the morphisms $F(p_I)$. Thus

$$\left(R(p_I): R(L) \longrightarrow RF(I)\right)_{I \in \mathcal{I}}$$

is the limit of the functor RF. $\qquad\qquad\qquad\qquad\qquad\qquad\qquad\qquad\qquad\qquad$ $\square$

5.2 Monads

A monad on a category C is an endofunctor on C satisfying "monoid-like axioms" relative to the composition of endomorphisms. More precisely:

Definition 5.7 A *monad* $\mathbb{T}$ on a category C is a triple (T, ε, μ) where

$$T: C \longrightarrow C$$

is a functor and

$$\varepsilon: \mathrm{id}_C \Rightarrow T, \quad \mu: T \circ T \Rightarrow T$$

are natural transformations making the following diagrams commute:

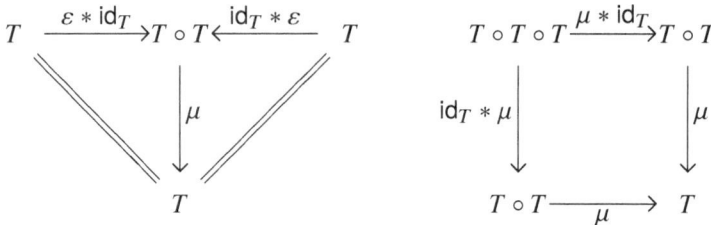

The natural transformation ε is called the *unit* of the monad and the natural transformation μ is called the *multiplication* of the monad

The dual notion is that of a *comonad*: a comonad on a category C is thus a monad on the dual category C^{op}. Trivially, the comonads will satisfy all the dual properties of those of a monad.

Here is a classical way to construct a monad from a given adjunction:

Proposition 5.8 *Let*

$$R: \mathcal{B} \longrightarrow \mathcal{A}, \quad L: \mathcal{A} \longrightarrow \mathcal{B}, \quad L \dashv R$$

be an adjunction, with corresponding unit and co-unit

$$\varepsilon: \mathrm{id}_{\mathcal{A}} \Rightarrow R \circ L, \quad \eta: L \circ R \Rightarrow \mathrm{id}_{\mathcal{B}}$$

(see Proposition 5.4). The functor

$$T = R \circ L: \mathcal{A} \longrightarrow \mathcal{A}$$

together with the natural transformations

$$\varepsilon: \mathrm{id}_{\mathcal{A}} \Rightarrow T, \quad \mu = \mathrm{id}_R * \eta * \mathrm{id}_L: T \circ T \Rightarrow T$$

constitute a monad $\mathbb{T} = (T, \varepsilon, \mu)$ *on* $\mathcal{A}$.

Proof The naturality of η implies, for each object $B \in \mathcal{B}$, the commutativity of the square

$$\begin{array}{ccc} LRLR(B) & \xrightarrow{\eta_{LR(B)}} & LR(B) \\ \scriptstyle{LR(\eta_B)} \downarrow & & \downarrow \scriptstyle{\eta_B} \\ LR(B) & \xrightarrow{\eta_B} & B \end{array}$$

Using the "triangular identities of the adjunction" (see Proposition 5.4), we get

$$\mu \circ (\varepsilon * 1_T) = (\mathrm{id}_R * \eta * \mathrm{id}_L) \circ (\varepsilon * \mathrm{id}_R * \mathrm{id}_L)$$
$$= \big((\mathrm{id}_R * \eta) \circ (\varepsilon * \mathrm{id}_R)\big) * \mathrm{id}_L$$
$$= \mathrm{id}_R * \mathrm{id}_L$$
$$= \mathrm{id}_T,$$
$$\mu \circ (\mathrm{id}_T * \varepsilon) = (\mathrm{id}_R * \eta * \mathrm{id}_L) \circ (\mathrm{id}_R * \mathrm{id}_L * \varepsilon)$$
$$= \mathrm{id}_R * \big((\eta * \mathrm{id}_L) \circ (\mathrm{id}_L * \varepsilon)\big)$$
$$= \mathrm{id}_R * \mathrm{id}_L$$
$$= \mathrm{id}_T,$$
$$\mu \circ (\mu * \mathrm{id}_T) = (\mathrm{id}_R * \eta * \mathrm{id}_L) \circ (\mathrm{id}_R * \eta * \mathrm{id}_L * \mathrm{id}_R * \mathrm{id}_L)$$
$$= \mathrm{id}_R * \big(\eta \circ (\eta * \mathrm{id}_L * \mathrm{id}_R)\big) * \mathrm{id}_L$$
$$= \mathrm{id}_R * \big(\eta \circ (\mathrm{id}_L * \mathrm{id}_R * \eta)\big) * \mathrm{id}_L$$
$$= (\mathrm{id}_R * \eta * \mathrm{id}_L) \circ (\mathrm{id}_R * \mathrm{id}_L * \mathrm{id}_R * \eta * \mathrm{id}_L)$$
$$= \mu \circ (\mathrm{id}_T * \mu),$$

and this forces the conclusion.   □

Conversely, each monad induces an adjunction.

Definition 5.9 Let $\mathbb{T} = (T, \varepsilon, \mu)$ be a monad on the category C.

1. A $\mathbb{T}$-algebra is a pair (C, ξ) where

$$C \in C, \quad \xi \colon T(C) \longrightarrow C$$

and the following squares are commutative

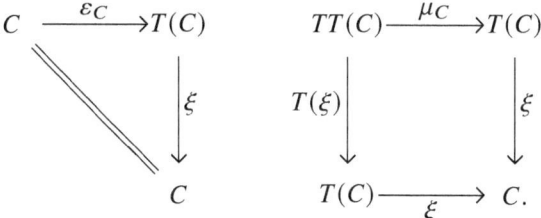

2. A morphism $f \colon (C, \xi) \longrightarrow (D, \zeta)$ of $\mathbb{T}$-algebras is a morphism $f \colon C \longrightarrow D$ in C making the following diagram commute:

$$\begin{array}{ccc} T(C) & \xrightarrow{\ T(f)\ } & T(D) \\ \downarrow{\scriptstyle \xi} & & \downarrow{\scriptstyle \zeta} \\ C & \xrightarrow{\ f\ } & D \end{array}$$

3. The category of $\mathbb{T}$-algebras is called the *Eilenberg–Moore category* of the monad $\mathbb{T}$ and is denoted by $C^{\mathbb{T}}$.

Proposition 5.10 *Let* $\mathbb{T} = (T, \varepsilon, \mu)$ *be a monad on the category* C. *The functor*

$$U^{\mathbb{T}}: C^{\mathbb{T}} \longrightarrow C, \quad (C, \xi) \mapsto C, \quad f \mapsto f$$

1. *is faithful;*
2. *reflects monomorphisms, epimorphisms and isomorphisms;*
3. *admits a left adjoint functor* $F^{\mathbb{T}}$:

$$F^{\mathbb{T}}: C \longrightarrow C^{\mathbb{T}}, \quad C \mapsto (T(C), \mu_C), \quad f \mapsto T(f).$$

The $\mathbb{T}$-algebra $F^{\mathbb{T}}(C)$ *is called the* free $\mathbb{T}$-algebra on C.

Proof The first statement is trivial. In the same way if

$$f: (C, \xi) \longrightarrow (D, \zeta)$$

is such that f is a monomorphism or an epimorphism in C, it is trivially a monomorphism or an epimorphism in $C^{\mathbb{T}}$.

Next consider $f: (C, \xi) \longrightarrow (D, \zeta)$ in $C^{\mathbb{T}}$ and assume that f is an isomorphism in C. One has

$$f \circ \xi \circ T(f^{-1}) = \zeta \circ T(f) \circ T(f^{-1}) = \zeta = f \circ f^{-1} \circ \zeta.$$

Thus $\xi \circ T(f^{-1}) = f^{-1} \circ \zeta$ because f is an isomorphism. This implies that the morphism $f^{-1}: (D, \zeta) \longrightarrow (C, \xi)$ is a morphism of $\mathbb{T}$-algebras, inverse of f.

Finally if $C \in C$, observe that the axioms for a monad imply immediately that $(T(C), \mu_C)$ is a $\mathbb{T}$-algebra. In the same way, given a morphism $s: C \longrightarrow C'$ in C,

$$T(s): (T(C), \mu_C) \longrightarrow (T(C'), \mu_{C'})$$

is a morphism of $\mathbb{T}$-algebras. This defines the functor $F^{\mathbb{T}}$.

Let us now prove that composition with $\varepsilon_C: C \longrightarrow T(C)$

$$C^{\mathbb{T}}(F^{\mathbb{T}}(C), (D, \zeta)) \longrightarrow C(C, U^{\mathbb{T}}(D, \zeta)), \quad g \mapsto g \circ \varepsilon_C$$

is a natural bijection. To have a bijection, it suffices to prove that for each morphism $h \in C(C, D)$ in C, there exists a unique morphism $g \in C^{\mathbb{T}}((TC, \mu_C), (D, \zeta))$ such that $g \circ \varepsilon_C = h$.

To prove the uniqueness of g, it suffices to observe that the monad and $\mathbb{T}$-algebra axioms imply

$$g = g \circ \mu_C \circ T(\varepsilon_C) = \zeta \circ T(g) \circ T(\varepsilon_C) = \zeta \circ T(h).$$

To prove the existence of g, it suffices now to check that $g = \zeta \circ T(h)$ does the job. Indeed, g is a morphism of $\mathbb{T}$-algebras because

$$g \circ \mu_C = \zeta \circ T(h) \circ \mu_C = \zeta \circ \mu_D \circ TT(h) = \zeta \circ T(\zeta) \circ TT(h) = \zeta \circ T(g).$$

Moreover,

$$g \circ \varepsilon_C = \zeta \circ T(h) \circ \varepsilon_C = \zeta \circ \varepsilon_D \circ h = h.$$

The naturality of the bijections is trivial. $\qquad\qquad\qquad\qquad\qquad\qquad\qquad\square$

Corollary 5.11 *Let $\mathbb{T} = (T, \varepsilon, \mu)$ be a monad on a category C. The monad $\mathbb{T}$ is also the monad associated with its Eilenberg–Moore adjunction of Proposition 5.8.*

Proof We use the notation of Proposition 5.10, whose proof implies at once that $U^{\mathrm{T}} \circ F^{\mathrm{T}} = T$ and

$$\varepsilon \colon \mathrm{id}_C \Rightarrow T = U^{\mathrm{T}} \circ F^{\mathrm{T}}$$

is the unit of the adjunction $F^{\mathrm{T}} \dashv U^{\mathrm{T}}$. Comparing with the construction in Proposition 5.8, it suffices now to check that $\mu = U^{\mathrm{T}} * \eta * F^{\mathrm{T}}$, where η is the co-unit of the adjunction $F^{\mathrm{T}} \dashv U^{\mathrm{T}}$.

If (X, ξ) is a $\mathbb{T}$-algebra, the co-unit

$$\eta_{(X,\xi)} \colon (F^{\mathrm{T}} \circ U^{\mathrm{T}})(X, \xi) \longrightarrow (X\xi)$$

is the unique morphism of C^{T} such that the triangle

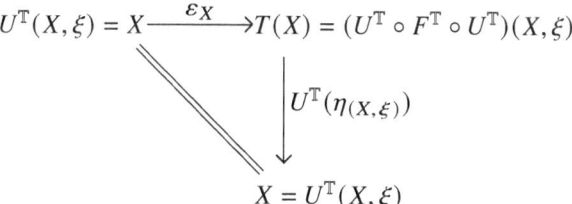

commutes; this is just the particular case $h = \mathrm{id}_X$ in the proof of Proposition 5.10. Thus the morphism $\eta_{(X,\xi)}$ is the corresponding morphism g, that is, $\eta_{(X,\xi)} = \xi$. Therefore for each $C \in C$,

$$U^{\mathrm{T}}\big(\eta_{F^{\mathrm{T}}(C)}\big) = U^{\mathrm{T}}\big(\eta_{(T(C),\mu_C)}\big) = \mu_C. \qquad\qquad\qquad\qquad\square$$

5.3 Monadic Categories

Applying the results of Section 5.2, we can construct a monad from a given adjunction and next, an Eilenberg–Moore adjunction from that monad. Let us now compare these two adjunctions.

Proposition 5.12 *Let L ⊣ R be adjoint functors.*

$$L: \mathcal{A} \longrightarrow \mathcal{B}, \quad R: \mathcal{B} \longrightarrow \mathcal{A}.$$

Let $\mathbb{T}$ *be the monad on* $\mathcal{A}$ *induced by that adjunction (see Proposition 5.8). Consider further the Eilenberg–Moore adjunction induced by the monad* $\mathbb{T}$ *(see Proposition 5.10). With the notation of Proposition 5.8, there exists a "comparison functor"*

$$J: \mathcal{B} \longrightarrow C^{\mathsf{T}}, \quad B \mapsto (R(B), R(\eta_B)), \quad b \mapsto R(b)$$

such that both triangles

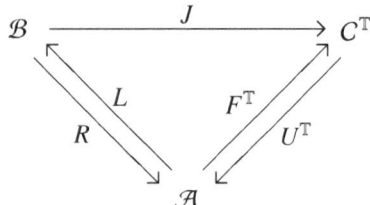

are commutative.

Proof First, observe that

$$R(\eta_B): TR(B) = RLR(B) \longrightarrow R(B)$$

defines a $\mathbb{T}$-algebra structure on $R(B)$:

$$
\begin{aligned}
R(\eta_B) \circ \varepsilon_{R(B)} &= R(\eta_B) \circ \alpha_{R(B)} \\
&= \mathsf{id}_{R(B)}, \\
R(\eta_B) \circ \mu_{R(B)} &= R(\eta_B) \circ R(\eta_{LR(B)}) \\
&= R(\eta_B \circ \eta_{LR(B)}) \\
&= R(\eta_B \circ LR(\eta_B)) \\
&= R(\eta_B) \circ T(R(\eta_B)).
\end{aligned}
$$

Next, each morphism $b: B \longrightarrow C$ of $\mathcal{B}$ induces a morphism

$$R(b): (R(B), R(\eta_B)) \longrightarrow (R(C), R(\eta_C))$$

of $\mathbb{T}$-algebras, because

$$
\begin{aligned}
R(\eta_C) \circ TR(b) &= R(\eta_C) \circ RLR(b) \\
&= R(\eta_C \circ LR(b)) \\
&= R(b \circ \eta_B) \\
&= R(b) \circ R(\eta_B).
\end{aligned}
$$

Clearly, J is a functor such that $U \circ J \cong R$. Moreover, Propositions 5.8 and 5.10 imply

$$(J \circ L)(A) = \big(RL(A), R(\eta_{L(A)})\big) = (T(A), \mu_A) = F^{\mathrm{T}}(A)$$

for each object $A \in \mathcal{A}$, and in the same way for each morphism of $\mathcal{A}$.   □

Definition 5.13 A functor $R \colon \mathcal{B} \longrightarrow \mathcal{A}$ is monadic when

1. R admits a left adjoint functor L;
2. if $\mathbb{T}$ is the monad associated with the adjunction $L \dashv R$ (see Proposition 5.8), the comparison functor $J \colon \mathcal{B} \longrightarrow C^{\mathrm{T}}$ as in Proposition 5.12 is an equivalence of categories.

In that situation, the category $\mathcal{B}$ is also said to be *monadic* over $\mathcal{A}$.

5.4 The Beck Monadicity Criterion

Beck's criterion characterizes monadic functors. To state it, we first need some terminology.

Definition 5.14 A diagram

$$C \underset{v}{\overset{u}{\rightrightarrows}} D \xrightarrow{q} Q$$

in a category X is called a *split coequalizer* when

$$q \circ u = q \circ v, \quad q \circ s = \mathrm{id}_Q, \quad u \circ r = \mathrm{id}_D, \quad v \circ r = s \circ q.$$

The following lemma justifies this terminology:

Lemma 5.15 *With the notation of Definition 5.14, $q = \mathrm{Coker}(u, v)$ and this coequalizer is preserved by every functor $F \colon X \longrightarrow \mathcal{Y}$ with domain X.*

Proof By assumption, $q \circ u = q \circ v$. If $p \colon D \longrightarrow P$ is another morphism such that $p \circ u = p \circ v$, we get

$$p \circ s \circ q = p \circ v \circ r = p \circ u \circ r = p \circ \mathrm{id}_D = p.$$

Thus $p \circ s$ is a factorization of p through q. This factorization is unique because q is an epimorphism, since it admits the section s.

Applying a functor F yields an analogous situation in $\mathcal{Y}$ and thus, the same argument implies that $F(q) = \mathrm{Coker}\big(F(u), F(v)\big)$.   □

Lemma 5.16 *Let* $\mathbb{T} = (T, \varepsilon, \mu)$ *be a monad on a category* C. *For each* $\mathbb{T}$-*algebra* (C, ξ), *the following diagram is a coequalizer in* $C^{\mathbb{T}}$

$$\left(TT(C), \mu_{T(C)}\right) \underset{T(\xi)}{\overset{\mu_C}{\rightrightarrows}} \left(T(C), \mu_C\right) \overset{\xi}{\longrightarrow} (C, \xi)$$

and its image by the functor $U^{\mathbb{T}} : C^{\mathbb{T}} \longrightarrow C$ *yields a split coequalizer in* C.

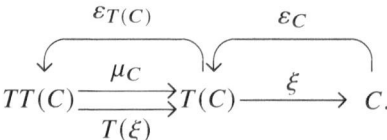

$$TT(C) \underset{T(\xi)}{\overset{\mu_C}{\rightrightarrows}} T(C) \overset{\xi}{\longrightarrow} C.$$

Proof In C, the equalities $\xi \circ \mu_C = \xi \circ T(\xi)$ and $\xi \circ \varepsilon_C = \mathrm{id}_C$ hold by definition of a $\mathbb{T}$-algebra (see Definition 5.9). The equality $\mu_C \circ \varepsilon_{T(C)} = \mathrm{id}_{T(C)}$ holds by definition of a monad (see Definition 5.7). Finally the equality $T(\xi) \circ \varepsilon_{T(C)} = \varepsilon_C \circ \xi$ holds by naturality of ε. Thus in C, we have indeed the split coequalizer announced in the statement.

It remains to prove that we also have the announced coequalizer in $C^{\mathbb{T}}$. By Proposition 5.10, we know that

$$F^{\mathbb{T}}(T(C)) = \left(TT(C), \mu_{T(C)}\right) \quad \text{and}$$
$$F^{\mathbb{T}}(C) = \left(T(C), \mu_C\right)$$

are $\mathbb{T}$-algebras. The morphism μ_C is a morphism of $\mathbb{T}$-algebras, by definition of a monad (see Definition 5.7). The morphism $T(\xi)$ is a morphism of $\mathbb{T}$-algebras by naturality of μ. Finally the morphism ξ is a morphism of $\mathbb{T}$-algebras by definition of a $\mathbb{T}$-algebra (see Definition 5.9).

The morphism

$$F^{\mathbb{T}}(\varepsilon_C) = T(\varepsilon_C) : F^{\mathbb{T}}(C) \longrightarrow F^{\mathbb{T}}(T(C))$$

is a section of both morphisms μ_C and $T(\xi)$. Indeed, in C we have

$$\mu_C \circ T(\varepsilon_C) = \mathrm{id}_{T(C)},$$
$$T(\xi) \circ T(\varepsilon_C) = \mathrm{id}_{T(C)},$$

by definitions of a monad and a $\mathbb{T}$-algebra. We know already that $\xi \circ \mu_C = \xi \circ T(\xi)$. Moreover, if $f : (T(C), \mu_C) \longrightarrow (D, \zeta)$ is a morphism in $C^{\mathbb{T}}$ such that $f \circ \mu_C = f \circ T(\xi)$, the composite

$$C \overset{\varepsilon_C}{\longrightarrow} T(C) \overset{f}{\longrightarrow} D$$

is a factorization of f through ξ in C (see Lemma 5.15). It suffices now to prove that $f \circ \varepsilon_C : (C, \xi) \longrightarrow (D, \zeta)$ is a morphism of $\mathbb{T}$-algebras. Indeed,

$$\zeta \circ T(f) \circ T(\varepsilon_C) = f \circ \mu_C \circ T(\varepsilon_C)$$
$$= f$$
$$= f \circ \mu_C \circ \varepsilon_{T(C)}$$
$$= f \circ T(\xi) \circ \varepsilon_{T(C)}$$
$$= f \circ \varepsilon_C \circ \xi,$$

The uniqueness holds because it does in C and U^T is faithful (see Proposition 5.10).□

We are now ready to prove Beck's criterion.

Theorem 5.17 (Beck's criterion) *Let* $R\colon X \longrightarrow C$ *be a functor. The following conditions are equivalent:*

1. *the functor R is monadic;*
2. a. *the functor R has a left adjoint L;*
   b. *the functor R reflects isomorphisms;*
   c. *if a pair $u, v\colon X \overrightarrow{\longrightarrow} Y$ in X is such that $(R(u), R(v))$ admits a split coequalizer in C, then (u, v) admits a coequalizer in X which is preserved by R.*

Proof $(1 \Rightarrow 2)$. By Proposition 5.10, it remains to consider a pair of morphisms $u, v\colon (C, \xi) \overrightarrow{\longrightarrow} (D, \zeta)$ in C^T and morphisms q, r, s in C which exhibit a split coequalizer in C (see Definition 5.14).

Consider the composite

$$\rho\colon T(Q) \xrightarrow{\ T(s)\ } T(D) \xrightarrow{\ \zeta\ } D \xrightarrow{\ q\ } Q$$

and let us prove that

$$(C, \xi) \underset{v}{\overset{u}{\rightrightarrows}} (D, \zeta) \xrightarrow{\ q\ } (Q, \rho)$$

is a coequalizer in C^T. Of course, this coequalizer will be mapped by U^T to the coequalizer $q = \mathsf{Coker}(u, v)$ in C.

First, (Q, ρ) is a $\mathbb{T}$-algebra:

$$\rho \circ \varepsilon_Q = q \circ \zeta \circ T(s) \circ \varepsilon_Q$$
$$= q \circ \zeta \circ \varepsilon_D \circ s$$
$$= q \circ s$$
$$= \mathsf{id}_Q,$$

$$\rho \circ T(\rho) = q \circ \zeta \circ T(s) \circ T(q) \circ T(\zeta) \circ TT(s)$$
$$= q \circ \zeta \circ T(v) \circ T(r) \circ T(\zeta) \circ TT(s)$$
$$= q \circ v \circ \xi \circ T(r) \circ T(\zeta) \circ TT(s)$$
$$= q \circ u \circ \xi \circ T(r) \circ T(\zeta) \circ TT(s)$$
$$= q \circ \zeta \circ T(u) \circ T(r) \circ T(\zeta) \circ TT(s)$$
$$= q \circ \zeta \circ T(\zeta) \circ TT(s)$$
$$= q \circ \zeta \circ \mu_D \circ TT(s)$$
$$= q \circ \zeta \circ T(s) \circ \mu_Q$$
$$= \rho \circ \mu_Q.$$

Second, $q \colon (D, \zeta) \longrightarrow (Q, \rho)$ is a morphism of $\mathbb{T}$-algebras:

$$\rho \circ T(q) = q \circ \zeta \circ T(s) \circ T(q)$$
$$= q \circ \zeta \circ T(v) \circ T(r)$$
$$= q \circ v \circ \xi \circ T(r)$$
$$= q \circ u \circ \xi \circ T(r)$$
$$= q \circ \zeta \circ T(u) \circ T(r)$$
$$= q \circ \zeta.$$

Finally if $p \colon (D, \zeta) \longrightarrow (P, \tau)$ in $C^{\mathbb{T}}$ is such that $p \circ u = p \circ v$, we get a factorization $w \colon Q \longrightarrow P$ in C such that $p = w \circ q$, because $q = \mathsf{Coker}(u, v)$ in C. It suffices to check that $w \colon (Q, \rho) \longrightarrow (P, \tau)$ is a morphism of $\mathbb{T}$-algebras. Indeed

$$\tau \circ T(w) = \tau \circ T(w) \circ T(q) \circ T(s)$$
$$= \tau \circ T(p) \circ T(s)$$
$$= p \circ \zeta \circ T(s)$$
$$= w \circ q \circ \zeta \circ T(s)$$
$$= w \circ \rho \circ T(q) \circ T(s)$$
$$= w \circ \rho.$$

To prove $(2 \Rightarrow 1)$, we refer to the characterization of equivalences in Lemma 2.36. Let $\mathbb{T}$ be the monad on C generated by the adjunction $L \dashv R$ (see Proposition 5.8). Let

$$J \colon X \longrightarrow C^{\mathbb{T}}, \quad X \mapsto \big(R(X), R(\eta_X)\big), \quad x \mapsto R(x)$$

be the comparison functor of Proposition 5.12. We must prove that J is an equivalence of categories (see Definition 5.13).

Consider first the pair of morphisms

$$\eta_{LR(X)},\ LR(\eta_X):\ LRLR(X) \underrightarrow{\hspace{2cm}} LR(X)$$

in X and their images under the functor R (see Propositions 5.8 and 5.10)

$$R(\eta_{LR(X)}) = \mu_{R(X)},\ \ RLR(\eta_X) = TR(\eta_X).$$

In C^T, we have the situation

$$\big(TTR(X),\mu_{R(X)}\big) \overset{\mu_{TR(X)}}{\underset{TR(\eta_X)}{\rightrightarrows}} \big(TR(X),\mu_X\big) \xrightarrow{\ R(\eta_X)\ } \big(R(X),R(\eta_X)\big) = J(X)$$

as described in Lemma 5.16. Thus the image of the pair $\big(\mu_{R(X)},TR(\eta_X)\big)$ by the functor U^T is equal to the image of the pair $\big(\eta_{LR(X)},LR(\eta_X)\big)$ by the functor R. By Lemma 5.16, we obtain so a split coequalizer in C. The last assumption in the statement implies now that the pair $\big(\eta_{LR(X)},LR(\eta_X)\big)$ of morphisms in X admits a coequalizer q preserved by the functor R.

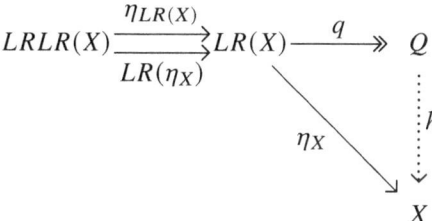

By naturality of η, we have

$$\eta_X \circ \eta_{LR(X)} = \eta_X \circ LR(\eta_X);$$

and this implies the existence of a factorization h through the coequalizer q. We know that the coequalizer q is preserved by the functor R and we know also that in C, the split coequalizer is $R(\eta_X)$. The factorization $R(h)$ is thus an isomorphism. But by assumption, the functor R reflects isomorphisms; thus h is an isomorphism in X. This proves that

$$\eta_X = \mathsf{Coker}\big(\eta_{LR(X)},LR(\eta_X)\big)$$

in X. Going back to the definition of J in Proposition 5.12, we conclude that the arguments above also prove that J preserves this coequalizer, because

$$J(\eta_X) = R(\eta_X) = \mathsf{Coker}\big(\mu_{R(X)},TR(\eta_X)\big) = \mathsf{Coker}\big(J(\eta_{LR(X)}),JLR(\eta_X)\big).$$

Let us now prove that J is full and faithful. For this, consider the diagram

$$
\begin{array}{ccc}
X(X,Y) & \xrightarrow{\ \ J_{X,Y}\ \ } & C^{\mathrm{T}}(J(X),J(Y)) \\
\downarrow{\scriptstyle X(\eta_X,Y)} & & \downarrow{\scriptstyle C^{\mathrm{T}}(J(\eta_X),J(Y))} \\
X(LR(X),Y) & \xrightarrow{\ \ J_{LR(X),Y}\ \ } & C^{\mathrm{T}}(JLR(X),J(Y)) \\
\downdownarrows{\scriptstyle X(LR(\eta_X),Y)\ \ X(\eta_{LR(X)},Y)} & & \downdownarrows{\scriptstyle C^{\mathrm{T}}(JLR(\eta_X),J(Y))\ \ C^{\mathrm{T}}(J(\eta_{LR(X)}),J(Y))} \\
X(LRLR(X),Y) & \xrightarrow[\ \ J_{LRLR(X),Y}\ \ ]{} & C^{\mathrm{T}}(JLRLR(X),J(Y))
\end{array}
$$

in the category Set of sets, with $X, Y \in X$. The horizontal mappings are induced by the action of J. We must prove that $J_{X,Y}$ is bijective. But

$$
\eta_X = \mathsf{Coker}\big(\eta_{LR(X)}, LR(\eta_X)\big)
$$

and we know already that the functor J preserves this coequalizer. Since both functors $X(-,Y)$ and $C^{\mathrm{T}}(-,J(Y))$ trivially transform coequalizers into equalizers (they are so-called "representable functors") the vertical lines of the diagram are equalizers. Therefore to prove that $J_{X,Y}$ – the factorization between the equalizers – is bijective, it suffices to prove that both other lines are bijective. Using both adjunctions $L \dashv R$ and $F^{\mathrm{T}} \dashv U^{\mathrm{T}}$, we get

$$
\begin{aligned}
X(LR(X),Y) &\cong C(R(X),R(Y)) \\
&\cong C\big(R(X),U^{\mathrm{T}}(R(Y),R(\eta_Y))\big) \\
&\cong C^{\mathrm{T}}\big(F^{\mathrm{T}}(R(X)),(R(Y),R(\eta_Y))\big) \\
&\cong C^{\mathrm{T}}\big((TR(X),\mu_{R(X)}),(R(Y),R(\eta_Y))\big) \\
&\cong C^{\mathrm{T}}\big((RLR(X),R(\eta_{LR(X)})),(R(Y),R(\eta_Y))\big) \\
&\cong C^{\mathrm{T}}\big(JLR(X),J(Y)\big).
\end{aligned}
$$

Thus the central horizontal line is bijective. Putting $X' = LR(X)$, this also implies that the bottom horizontal line is bijective.

Finally, we must prove that J is essentially surjective on the objects. Given a $\mathbb{T}$-algebra (C,ξ), consider the pair

$$
\eta_{L(C)}, L(\xi) \colon LRL(C) \rightrightarrows L(C)
$$

in X. Its image under the functor J is the pair of Lemma 5.16. The image in X of this pair under the functor R is also the image under U^{T} of the corresponding pair in

C^T, because $U^T \circ J = R$ (see Proposition 5.8). By Lemma 5.16, this pair in C admits a split coequalizer. Thus by assumption, there exists a coequalizer in X

$$LRL(C) \overset{\eta_{L(C)}}{\underset{L(\xi)}{\rightrightarrows}} L(C) \overset{p}{\twoheadrightarrow} P$$

such that, up to an isomorphism, $R(P) = C$ and $R(p) = \xi$. In particular, $R(p)$ admits a section σ and $LR(p)$ admits the section $L(\sigma)$; moreover $LR(p)$ is an isomorphism in X. By definition of p and by naturality of η,

$$p \circ LR(p) = p \circ L(\xi) = p \circ \eta_{LR(P)} = \eta_P \circ LR(p),$$

thus $p = \eta_P$, because $LR(p)$ is an epimorphism. Finally we get

$$J(P) = (R(P), R(\eta_P)) \cong (C, R(p)) = (C, \xi)$$

and this concludes the proof. □

Chapter 6
Profinite Groupoids and Presheaves

Abstract A groupoid is a category in which every morphism is an isomorphism. A group G determines a groupoid $\mathbb{G}$ with a single formal object $\star$, and $\mathbb{G}(\star, \star) = G$ as set of (iso)morphisms. The Galois theory of rings will use a Galois groupoid, with possibly several objects, instead of a group. A profinite groupoid will be one whose set of objects and set of morphisms are profinite spaces, while all operations are continuous. The notion of profinite presheaf on a profinite groupoid then generalizes the notion of profinite G-space, for a profinite group G. These notions and results in the profinite case are special instances of internal category theory in a category with pullbacks.

6.1 Profinite Groupoids

Let us recall that an arrow $f\colon A \longrightarrow B$ in a category is an *isomorphism* when there exists a morphism $g\colon B \longrightarrow A$ such that $g \circ f = \mathrm{id}_A$ and $f \circ g = \mathrm{id}_B$. Such a morphism g is necessarily unique and is generally written f^{-1}.

Definition 6.1 A groupoid is a small category in which every morphism is an iso-morphism.

To shorten the language, in this book, all groupoids will thus be small (see Definition 3.1). For example, from every small category C, one gets a groupoid when keeping the same objects but choosing as morphisms the isomorphisms of C. But the following example will be much more important in this book.

Proposition 6.2 *The category of groups is isomorphic to the category of groupoids with a single object.*

Proof If $\mathbb{G}$ is a groupoid with a single object $\star$, then its set $\mathbb{G}(\star, \star)$ of morphisms is a group G for the composition of morphisms. Conversely if G is a group, consider the groupoid $\mathbb{G}$ with a single object $\star$ and $\mathbb{G}(\star, \star) = G$ as set of morphisms; the

F. Borceux, *Galois Theories of Fields and Rings*, Coimbra Mathematical Texts 2,
https://doi.org/10.1007/978-3-031-58460-2_7

composition of arrows is the multiplication of G. This yields at once a bijection between groups and groupoids with a single object.

Doing the same with a group G' and the corresponding groupoid $\mathbb{G}'$, it is immediate that the group homomorphisms $f \colon G \longrightarrow G'$ are now in bijection with the functors $F \colon \mathbb{G} \longrightarrow \mathbb{G}'$. $\square$

We can thus view a groupoid as a generalization of the notion of group: a kind of group "with several objects".

Definition 6.3 A *profinite groupoid* is a groupoid $\mathbb{G}$ whose sets G_0 of objects and G_1 of all morphisms are provided with profinite topologies, in such a way that the following mappings are continuous:

- $d_0 \colon G_1 \longrightarrow G_0$, $d_0(f \colon A \to B) = A$;
- $d_1 \colon G_1 \longrightarrow G_0$; $d_1(f \colon A \to B) = B$;
- $n \colon G_0 \longrightarrow G_1$; $n(A) = \mathrm{id}_A$;
- $s \colon G_1 \longrightarrow G_1$; $s(f) = f^{-1}$;
- $m \colon G_2 = G_1 \times_{G_0} G_1 \longrightarrow G_1$; $m(g, f) = g \circ f$;

where G_2 is thus the pullback of d_0 and d_1, that is, the set of composable pairs of arrows, provided with the pullback topology: the topology induced by the product topology on $G_1 \times G_1$.

Observe that in this definition, the topology on G_2 is profinite as well, by Theorem 3.18 and Corollary 3.13.

And now as expected:

Proposition 6.4 *The category of profinite groups is isomorphic to the category of profinite groupoids with a single object.*

Proof With the notation of Definition 6.3, the continuity of d_0, d_1 and n is trivial when G_0 is a singleton. In that case G_2 is simply $G_0 \times G_0$ and the continuity of m is the continuity of the multiplication of the corresponding group, as in Proposition 6.2. The continuity of s reduces to the continuity of the inverse operation in G. $\square$

6.2 Profinite $\mathbb{G}$-Presheaves

We want now to generalize, to the case of a groupoid $\mathbb{G}$, the notion of G-set for a group G. And analogously in the profinite case.

Proposition 6.5 *Let G be a group and $\mathbb{G}$ the corresponding groupoid with a single object (see Proposition 6.2). There is an isomorphism of categories between the category of G-sets and the category of functors from $\mathbb{G}$ to the category Set of sets.*

Proof A functor $F \colon \mathbb{G} \longrightarrow \mathsf{Set}$ is determined by the single set $F(\star)$ and for each $g \in G$, the mapping

$$F(g): F(\star) \longrightarrow F(\star) \quad x \mapsto F(g)(x),$$

with the properties

$$F(1)(x) = x, \quad F(g \circ g')(x) = F(g)\big(F(g')(x)\big).$$

This means exactly that $F(\star)$, provided with the action

$$G \times F(\star) \longrightarrow F(\star), \quad (g, x) \mapsto F(g)(x),$$

is a G-set.

In exactly the same way, given another functor H, a natural transformation $F \Rightarrow H$ reduces to a single mapping $F(\star) \longrightarrow H(\star)$ commuting with the G-actions.   □

Proposition 6.5 indicates at once how to generalize the notion of G-set:

Definition 6.6 Let 𝔾 be a groupoid. By a 𝔾-presheaf[1] we mean a functor from 𝔾 to the category of sets. A morphism of 𝔾-presheaves is a natural transformation.

A warning is useful here: most often in the literature, "presheaf" refers to a "contravariant presheaf", while in the covariant case, one says explicitly "covariant presheaf". Since in this book we shall only consider covariant presheaves, to avoid an overly heavy terminology, we shall just call them "presheaves".

Our goal is now to define a profinite presheaf on a profinite groupoid 𝔾: this requires more attention than just considering a functor. With the notation of Definition 6.3, one wants of course to consider a functor from 𝔾 to the category Prof of profinite spaces. But such a functor $F : 𝔾 \longrightarrow$ Prof exhibits in particular the family $\big(F(A)\big)_{A \in G_0}$ of profinite spaces, indexed by the set G_0 of objects of 𝔾. However, G_0 is also provided with a profinite topology, and thus one would like the family $\big(F(A)\big)_{A \in G_0}$ to be "continuously indexed" by the space G_0. What can this possibly mean? The following lemma is the key to that more elaborate notion.

Definition 6.7 Given an arbitrary category C and an object $I \in C$, the category C/I

- has for objects the pairs (A, α) where A is an object of C and $\alpha : A \longrightarrow I$ is a morphism in C;
- a morphism $f : (A, \alpha) \longrightarrow (B, \beta)$ is a morphism $f : A \longrightarrow B$ in C such that $\beta \circ f = \alpha$.

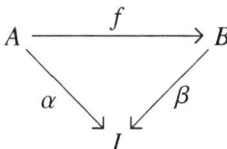

---

[1] The more involved notion of *sheaf* is relative to an additional structure put on 𝔾 and which does not appear in our context.

Lemma 6.8 *Given a set I, the category of I-families of sets is equivalent to the category* Set/I.

Proof The pair (A, α) determines an I-indexed family $\left(\alpha^{-1}(i)\right)_{i \in I}$ of subsets of A. Conversely, given a family $(A_i)_{i \in I}$ of sets, one can define $A = \amalg_{i \in I} A_i$, the disjoint union of these sets, and put $\alpha(a) = i$ when $a \in A_i$. Those two processes are trivially mutually inverse.

Moreover an arrow $f: (A, \alpha) \longrightarrow (B, \beta)$ is precisely an arrow $f: A \longrightarrow B$ whose restrictions yield a family of arrows $\left(f_i: \alpha^{-1}(i) \longrightarrow \beta^{-1}(i)\right)_{i \in I}$. $\square$

This suggests to define:

Definition 6.9 Let I be a profinite space. By a continuously I-indexed family of profinite spaces we mean a pair (A, α) where A is a profinite space and $\alpha: A \longrightarrow I$ is a continuous mapping.

Lemma 6.10 *A continuously indexed family (A, α) as in Definition 6.9 generates a set-theoretically I-indexed family of profinite spaces $\left(\alpha^{-1}(i)\right)_{i \in I}$.*

Proof Each singleton $\{i\}$ is closed in I because I is a Hausdorff space. Thus each $\alpha^{-1}(i)$ is closed in the profinite space A, thus is profinite (see Proposition 3.11 and Theorem 3.18). $\square$

Let us make clear that Lemma 6.10, in contrast to Lemma 6.8, does not exhibit an equivalence of categories. The set-indexed family $\left(\alpha^{-1}(i)\right)_{i \in I}$ of profinite spaces allows us to reconstruct the set A, but the topology of A is not determined by just the topologies of the subspaces $\alpha^{-1}(i)$.

We are ready to define the profinite presheaves:

Definition 6.11 Let $\mathbb{G}$ be a profinite groupoid. A profinite $\mathbb{G}$-presheaf is a triple (P, π, ξ) where

- $\pi: P \longrightarrow G_0$ is a continuously G_0-indexed family of profinite spaces, that is, a morphism in Prof;
- $\xi: G_1 \times_{G_0} P \longrightarrow P$ is a continuous action, where the left-hand pullback is that of d_0 and π in the category of profinite spaces;

with the property that the following data define a functor $F: \mathbb{G} \longrightarrow$ Prof:

- for each object $A \in G_0$, $F(A) = \pi^{-1}(A)$;
- given $f: B \to C$ in G_1 and $x \in F(B)$, $F(f)(x) = \xi(f, x) \in F(C)$.

It remains to define the morphisms of profinite presheaves, which of course must be natural transformations "acting continuously".

Definition 6.12 Let G be a profinite groupoid and (P, π, ξ), (Q, σ, ζ) two profinite $\mathbb{G}$-presheaves. A morphism $\alpha: (P, \pi, \xi) \Rightarrow (Q, \sigma, \zeta)$ of profinite $\mathbb{G}$-presheaves is a continuous mapping $\alpha: P \longrightarrow Q$ making commutative the triangle

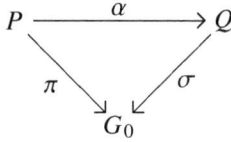

and such that its restrictions

$$\alpha_A\colon F(A) = \pi^{-1}(A)\longrightarrow\sigma^{-1}(A) = H(A)$$

for all $A \in G_0$ constitute a natural transformation between the corresponding functors F and H.

We shall write $\mathbb{G}$-ProfPresh to indicate the category of profinite $\mathbb{G}$-presheaves on a profinite groupoid $\mathbb{G}$.

It is now routine to prove that:

Proposition 6.13 *Given a profinite group G, the category of profinite G-spaces is isomorphic to the category of profinite $\mathbb{G}$-presheaves on the corresponding profinite groupoid $\mathbb{G}$ with a single object.*

Proof Indeed G_0 is now a singleton and therefore, a continuously G_0-indexed family (P, π) of profinite spaces reduces to giving just the profinite space P. □

6.3 The Monadicity of $\mathbb{G}$-Presheaves

The following monadicity result will be fundamental in the development of the Galois theory of rings.

Theorem 6.14 *Let $\mathbb{G}$ be a profinite groupoid. The functor*

$$\mathbb{G}\text{-ProfPresh}\longrightarrow\text{Prof}/G_0; \quad (P, \pi, \xi) \mapsto (P, \pi)$$

is monadic.

Proof We could use the Beck criterion (see Theorem 5.17) to prove this result, but the Galois theory of rings will necessitate the precise knowledge of the corresponding monad. We therefore give a more "constructive" proof.

Consider the functor

$$d_0^*\colon \text{Prof}/G_0\longrightarrow\text{Prof}/G_1$$

acting by pullback along $d_0\colon G_1\longrightarrow G_0$, and the functor

$$\Sigma_{d_1}\colon \text{Prof}/G_1\longrightarrow\text{Prof}/G_0$$

of composition with the morphism $d_1 : G_1 \longrightarrow G_0$ (see Definition 6.3). The monad $\mathbb{T}$ of the statement will have as functor part the composite

$$T = \Sigma_{d_1} \circ d_0^* : \mathsf{Prof}/G_0 \longrightarrow \mathsf{Prof}/G_1 \longrightarrow \mathsf{Prof}/G_0.$$

More explicitly:

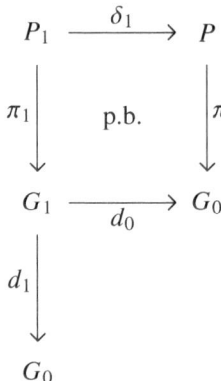

$$T(P, p) = (P_1, d_1 \circ \pi_1).$$

Let us observe that in this diagram, $P_1 = G_1 \times_{G_0} P$ is the pullback of d_0 and π, just as in Definition 6.11.

The unit

$$\varepsilon_{(P,\pi)} : (P, \pi) \longrightarrow T(P, \pi)$$

of the monad will be given by the mappings

$$\varepsilon_{(P,\pi)} : P \longrightarrow P_1, \quad (a \in \pi^{-1}(A)) \mapsto (\mathrm{id}_A, a).$$

These mappings are continuous as factorizations of the continuous mappings (see Definition 6.3)

$$\mathrm{id}_P : P = \!\!=\!\!= P, \quad n \circ p : P \longrightarrow G_1$$

through the pullback P_1. Moreover,

$$\left(d_1 \circ \pi_1 \circ \varepsilon_{(P,\pi)}\right)(a) = A = \pi(a),$$

thus $\varepsilon_{(P,p)}$ is a morphism in Prof/G_0. The naturality of ε is obvious.

To construct the multiplication of the monad, let us first compute $TT(P, \pi)$, which we shall abbreviate as $TT(P, \pi) = (P_2, d_1 \circ \pi_2)$. The object P_2 is the pullback of $(P_1, d_1 \circ p_1)$ along d_0:

$$P_2 \xrightarrow{\ \delta_2\ } P_1$$

$$\pi_2 \downarrow \qquad \text{p.b.} \qquad \downarrow d_1 \circ \pi_1$$

$$G_1 \xrightarrow{\ d_0\ } G_0$$

$$d_1 \downarrow$$

$$G_0$$

$$P_2 = \left\{ (h, (g, a)) \,\middle|\, (h \colon B \to C) \in G_1, \ (g, a) \in P_1, \ d_1(g) = d_0(h) \right\}$$
$$= \left\{ (h, g, a) \,\middle|\, g \colon A \to B, \ h \colon B \to C, \ a \in \pi^{-1}(A) \right\}.$$

Clearly, $\pi_2(h, (g, a)) = h$. The multiplication

$$\mu_{(P, p\pi)} \colon TT(P, \pi) \longrightarrow T(P, \pi)$$

of the monad is then given by the mappings

$$\mu_{(P, \pi)}(h, g, a) = (h \circ g, a).$$

Notice first that the mapping, with G_2 as defined in Definition 6.3,

$$\gamma \colon P_2 \longrightarrow G_2, \quad (h, g, a) \mapsto h \circ g$$

is continuous, because it is the factorization of the two continuous mappings

$$\pi_2 \colon P_2 \longrightarrow G_1, \quad \pi_1 \circ \delta_2 \colon P_2 \longrightarrow G_1$$

through the pullback G_2. Therefore the mapping $\mu_{(P, \pi)}$ is continuous as a factorization through the pullback P_1 of the two continuous mappings

$$m \circ \gamma \colon P_2 \longrightarrow G_1, \quad \delta_1 \circ \delta_2 \colon P_2 \longrightarrow P.$$

Moreover $\mu_{(P, p)}$ is a morphism of Prof/G_0 because

$$\left(d_1 \circ \pi_1 \circ \mu_{(P, \pi)} \right)(h, g, a) = d_1(h \circ g) = d_1(h) = (d_1 \circ \pi_2)(h, g, a).$$

Again, the naturality of μ is trivial.

With the notation above, the first two axioms for a monad mean

$$\mu_{(P, \pi)}(\mathsf{id}_B, g, a) = (g, a), \quad \mu_{(P, \pi)}(g, \mathsf{id}_A, a) = (g, a)$$

and they are trivially satisfied. The third axiom means that if $k: C \longrightarrow D$ is another morphism, then

$$((k \circ h) \circ g, a) = (k \circ (h \circ g), a),$$

which again, is trivial.

We have thus already constructed a monad $\mathbb{T} = (T, \varepsilon, \mu)$ on the category Prof/G_0. A $\mathbb{T}$-algebra (P, p, ξ) is an object $\pi: P \longrightarrow G_0$ of Prof/G_0, together with an action

$$\xi: T(P, \pi) = P_1 = G_1 \times_{G_0} P \longrightarrow P$$

such that, writing again $F(A) = \pi^{-1}(A)$,

1. for each $a \in F(A)$, $\xi(\mathsf{id}_A, a) = a$;
2. for each $(h, g, a) \in P_2$, $\xi(h \circ g, a) = \xi(h, \xi(g, a))$.

Using further the notation $\xi(g, a) = F(g)(a)$ these equalities become

1. $F(\mathsf{id}_A)(a) = a$;
2. $F(h \circ g)(a) = F(h)(F(g)(a))$.

These are the axioms for being a functor. Thus a $\mathbb{T}$-algebra is exactly a profinite G-presheaf.

Finally a morphism $\alpha: (P, \pi, \xi) \longrightarrow (Q, \sigma, \zeta)$ of $\mathbb{T}$-algebras is a continuous mapping $\alpha: P \longrightarrow Q$ over G_0 such that

$$(\alpha \circ \xi)(g, a) = (\zeta \circ T(\xi))(g, a).$$

But since α is a morphism of Prof/G_0,

$$a \in F(A) \Rightarrow \alpha(a) \in H(A),$$

where thus H is the functor determined by (Q, σ, ζ). Write $\alpha_A: F(A) \longrightarrow H(A)$ for the restriction of α. The $\mathbb{T}$-algebra axiom becomes, for each $g: A \longrightarrow B$ and $a \in F(A)$

$$\alpha_B(F(g)(a)) = H(g)(\alpha_A(a)).$$

This is exactly the condition for being a natural transformation. Thus a morphism of $\mathbb{T}$-algebras is exactly a morphism of profinite G-presheaves. □

Notice that in the proof of Theorem 6.14, we did not use the existence of inverses in the profinite groupoid G. Thus the same proof applies to the case of a profinite category G, instead of a profinite groupoid.

Chapter 7
The Descent Theory of Rings

Convention. *In this chapter, all rings and all algebras are commutative with unit.*

Abstract A morphism $\sigma\colon R \longrightarrow S$ of rings induces a pair of adjoint functors between the categories of R-modules and S-modules. Via this adjunction, the category of S-modules is always monadic over the category of R-modules: this implies that we can view an S-module as being an R-module with an additional structure. The morphism $\sigma\colon R \longrightarrow S$ of rings is a morphism of *effective descent* when, moreover, the category of R-modules is co-monadic over the category of S-modules; in that case, each R-module can thus also be seen as an S-module with an additional structure. We prove that the effective descent morphisms of rings are exactly the *pure* ones: the injective morphisms, which remain injective when tensored with whatever R-module. The descent theorem for rings implies an analogous result for algebras.

7.1 The Group $\mathbb{Q}/\mathbb{Z}$

Let us first recall two classical definitions:

Definition 7.1 An object C in a category C is a *cogenerator* when

$$\forall f, g\colon A \rightrightarrows B \quad (f = g) \Leftrightarrow (\forall h\colon B \longrightarrow C \ \ h \circ f = h \circ g).$$

Proposition 7.2 *Let C be an object in a category C. The following conditions are equivalent:*

1. *C is a cogenerator;*
2. *the representable functor*

$$C(-, C)\colon C \longrightarrow \mathsf{Set}, \quad D \mapsto C(D, C)$$

is faithful.

Proof Of course, $f = g$ always implies $h \circ f = h \circ g$. Thus condition 2 just rephrases condition 1.   □

For example in the category Set of sets, the set **2** $= \{0, 1\}$ is a cogenerator. Let us work by contraposition. With the notation of Definition 7.1, if $f \neq g$, it suffices to choose $a \in A$ such that $f(a) \neq g(a)$ and a mapping h such that $h\big(f(a)\big) = 0$, $h\big(g(a)\big) = 1$. More generally, every set with at least two elements is a cogenerator in the category Set.

Definition 7.3 An object C in a category C is injective when for each monomorphism s and each morphism f as in the diagram

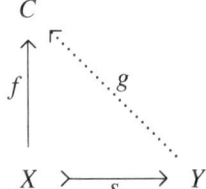

there exists a morphism g extending f, that is, $g \circ s = f$.

Let us insist on the fact that in this definition, injectivity requires the existence of g, not its uniqueness.

Proposition 7.4 *Let C be an object in a category C. The following conditions are equivalent:*

1. *C is an injective object;*
2. *the representable functor*

$$C(-, C)\colon C \longrightarrow \mathsf{Set}, \quad D \mapsto C(D, C)$$

transforms monomorphisms into epimorphisms.

Proof The epimorphisms in Set are the surjections. Thus condition 2 rephrases condition 1.   □

In the category Set of sets, **2** $= \{0, 1\}$ is also an injective object: with the notation of Proposition 7.4, it suffices to extend f by putting (for example) $f(y) = 0$ when $y \in Y \setminus X$. More generally, every non-empty set is injective in Set.

The existence of an injective cogenerator, like **2** in the category of sets, is generally a rather strong property for a category. This is in particular the case in the category of modules over a ring: a result that we shall prove in order to develop descent theory for rings.

Theorem 7.5 *The group $\mathbb{Q}/\mathbb{Z}$ is injective in the category Ab of abelian groups.*

Proof Consider an injection $s\colon A \rightarrowtail B$ of abelian groups and a group homomorphism $f\colon A \longrightarrow \mathbb{Q}/\mathbb{Z}$. Consider further all the intermediate extensions, that is, all the commutative diagrams in Ab

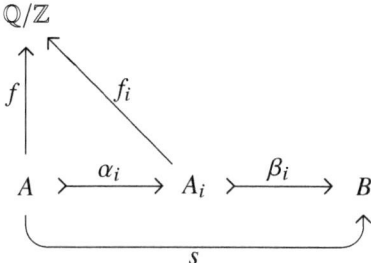

where α_i and β_i are inclusions of subgroups. The set of all these extensions is ordered by the relation

$$(A_i, f_i) \leq (A_j, f_j) \text{ when } A_i \subseteq A_j, \ f_j|_{A_i} = f_i.$$

If $(A_i, f_i)_{i \in I}$ is a chain of such intermediate extensions, $\overline{A} = \bigcup_{i \in I} A_i$ is still a subgroup of B and

$$\overline{f} : \overline{A} \longrightarrow \mathbb{Q}/\mathbb{Z}, \ b \mapsto f_i(b) \ \text{ if } b \in A_i$$

defines without ambiguity a new intermediate extension. Thus the intermediate extensions satisfy the assumptions of Zorn's lemma, already mentioned in Proposition 3.21, and therefore there exists a maximal intermediate extension (A', f'). We must now prove that $A' = B$.

If $A' \neq B$, consider the subgroup $A' + \mathbb{Z}b \subseteq B$, with $b \in B \setminus A'$. It suffices to prove the existence of an extension

$$\tilde{f} : A' + \mathbb{Z}b \longrightarrow \mathbb{Q}/\mathbb{Z}$$

of f' to get a contradiction, by maximality of (A', f'). If $A' \cap \mathbb{Z}b = (0)$, the sum is a direct one and it suffices to define $\tilde{f} = f'$ on A' and $\tilde{f}(zb) = 0$, for each $z \in \mathbb{Z}$, to obtain an extension of f'. And when $A' \cap \mathbb{Z}b \neq (0)$, consider

$$n = \min\{m | 0 \neq m \in \mathbb{N}, \ mb \in A'\}.$$

It suffices to define $\tilde{f} = f'$ on A' and, for each $z \in \mathbb{Z}$,

$$\tilde{f}(zb) = \left[\frac{z f'(nb)}{n} \right],$$

to get an extension of f'. This definition is unambiguous because, if $0 \neq m \in \mathbb{N}$ is such that $mb \in A'$, we can perform the Euclidean division of m by n

$$m = nq + r, \ r < n.$$

We obtain in particular

$$rb = mb - nqb \in A'$$

thus $r = 0$ by minimality of n. This implies $m = nq$ and

$$\tilde{f}(mb) = \tilde{f}(nqb) = \left[\frac{nqf'(nb)}{n}\right] = \left[qf'(nb)\right] = f'(qnb) = f'(mb).$$

It is trivial to check that $\tilde{f}$ is a group homomorphism. $\qquad\qquad\square$

Theorem 7.5 is a special case of a more general theorem which asserts that each divisible abelian group is injective.

Theorem 7.6 *The group* $\mathbb{Q}/\mathbb{Z}$ *is an injective cogenerator in the category* Ab *of abelian groups.*

Proof Consider two distinct group homomorphisms $f, g\colon A \rightrightarrows B$ and an element $a \in A$ such that $f(a) \neq g(a)$. Put $b = f(a) - g(a) \in B$.

If for each $0 \neq n \in \mathbb{N}$, $nb \neq 0$, the mapping

$$h\colon \mathbb{Z}b \longrightarrow \mathbb{Q}/\mathbb{Z}, \quad zb \mapsto z$$

is trivially a group homomorphism such that $h(b) = 1 \neq 0$.

If there exists $0 \neq m \in \mathbb{N}$ such that $mb = 0$, let us put

$$n = \min\{m \,|\, 0 \neq m \in \mathbb{N}, \ mb = 0\}.$$

In this case

$$h\colon \mathbb{Z}b \longrightarrow \mathbb{Q}/\mathbb{Z}, \quad zb \mapsto \left[\frac{z}{n}\right]$$

is a group homomorphism. Indeed, this definition is unambiguous because if $mb = m'b$, we can perform the Euclidean division of $m - m'$ by n:

$$m - m' = nq + r, \quad r < n.$$

We obtain

$$rb = (m - m')b - nqb \in A',$$

thus $r = 0$ by the minimality of n. This implies $m - m' = nq$ and

$$h(mb) = \left[\frac{m}{n}\right] = \left[\frac{m' + nq}{n}\right] = \left[\frac{m'}{n}\right] + \left[\frac{nq}{n}\right] = \left[\frac{m'}{n}\right] + \left[q\right] = \left[\frac{m'}{n}\right] = h(m'b)$$

because $[q] = 0 \in \mathbb{Q}/\mathbb{Z}$. This mapping h is trivially a group homomorphism such that $h(b) = [\frac{1}{n}] \neq 0$.

Thus in both cases we can now apply Theorem 7.5 to extend h as a group homomorphism

$$\overline{h}\colon B \longrightarrow \mathbb{Q}/\mathbb{Z}.$$

In particular $\overline{h}(b) = h(b) \neq 0$, thus $h(f(a) - g(a)) \neq 0$ and finally, $h(f(a)) \neq h(g(a))$. $\qquad\qquad\square$

We switch now to the case of modules over a ring.

Lemma 7.7 *Let R be a ring. The functor*

$$\mathsf{Ab}(-,\mathbb{Q}/\mathbb{Z}): \mathsf{Mod}_R \longrightarrow \mathsf{Ab}, \quad M \mapsto \mathsf{Ab}(M,\mathbb{Q}/\mathbb{Z})$$

factors through the category of R-modules.

Proof For each R-module M, if $r \in R$ and $f \in \mathsf{Ab}(M,\mathbb{Q}/\mathbb{Z})$, it suffices to define

$$rf: M \longrightarrow \mathbb{Q}/\mathbb{Z}, \quad m \mapsto f(rm)$$

to provide $\mathsf{Ab}(M,\mathbb{Q}/\mathbb{Z})$ with the structure of an R-module.
    If $g: M \longrightarrow M'$ is R-linear, $m \in M$ and $r \in R$,

$$\mathsf{Ab}(g,\mathsf{id})(rf)(m) = (rf \circ g)(m) = f\big((rg(m)\big) = r(f \circ g)(m) = r\mathsf{Ab}(g,\mathsf{id})(f)(m),$$

thus $\mathsf{Ab}(g,\mathsf{id})$ is R-linear.   □

Proposition 7.8 *Let R be a ring and Mod_R the category of R-modules. The functor*

$$\mathsf{Ab}(-,\mathbb{Q}/\mathbb{Z}): \mathsf{Mod}_R \longrightarrow \mathsf{Mod}_R, \quad M \mapsto \mathsf{Ab}(M,\mathbb{Q}/\mathbb{Z})$$

is exact and reflects isomorphisms.

Proof Limits and colimits of R-modules are computed as in the case of abelian groups. Just by definition of limit and colimit, the additive functor $\mathsf{Ab}(-,\mathbb{Q}/\mathbb{Z})$ transforms colimits into limits. In particular, it transforms cokernels into kernels. The functor $\mathsf{Ab}(-,\mathbb{Q}/\mathbb{Z})$ also transforms monomorphisms into epimorphisms, that is, injective linear mappings into surjective linear mappings. Indeed if $s: A \rightarrowtail B$ is an R-submodule, it is also a subgroup and $\mathsf{Ab}(s,\mathbb{Q}/\mathbb{Z})$ is surjective because $\mathbb{Q}/\mathbb{Z}$ is an injective object in Ab (see Theorem 7.5 and Proposition 7.4).
    This proves that the functor $\mathsf{Ab}(-,\mathbb{Q}/\mathbb{Z})$ is exact, that is, transforms a short exact sequence into a short exact sequence. Indeed

$$0 \longrightarrow A \xrightarrow{\ f\ } B \xrightarrow{\ g\ } C \longrightarrow 0$$

is a short exact sequence of R-modules as soon as f is a monomorphism and $g = \mathsf{Coker}\, f$, or equivalently, when g is an epimorphism and $f = \mathsf{Ker}\, g$.
    Finally, let $f: A \longrightarrow B$ be an R-linear mapping such that $\mathsf{Ab}(f,\mathsf{id}_{\mathbb{Q}/\mathbb{Z}})$ is an isomorphism. Consider the kernel k and the cokernel q of f.

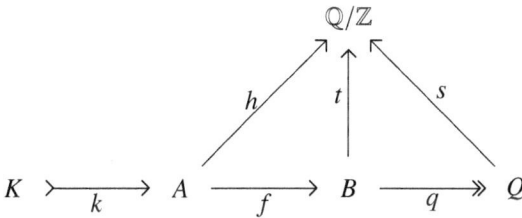

If $k \neq 0$, there exists a group homomorphism h such that $h \circ k \neq h \circ 0 = 0$, because $\mathbb{Q}/\mathbb{Z}$ is a cogenerator in Ab (see Theorem 7.6). But $\mathsf{Ab}(f, \mathrm{id}_{\mathbb{Q}/\mathbb{Z}})$ is an isomorphism; thus there exists a group homomorphism t such that $t \circ f = h$. This forces a contradiction:

$$0 \neq h \circ k = t \circ f \circ k = t \circ 0 = 0.$$

Thus necessarily, $k = 0$ and f is injective. In the same way if $q \neq 0$, there exists an s such that $s \circ q \neq 0$ and again we get a contradiction: $s \circ q \circ f = s \circ 0 = 0$, thus $s \circ q = 0$ because $\mathsf{Ab}(f, \mathbb{Q}/\mathbb{Z})$ is an isomorphism. This proves that $q = 0$ and f is surjective. $\square$

Lemma 7.9 *Let R be a ring and M, M' two R-modules. There exists an isomorphism of R-modules*

$$\theta_{M,M'} : \mathsf{Mod}_R\big(M, \mathsf{Ab}(M', \mathbb{Q}/\mathbb{Z})\big) \cong \mathsf{Ab}\big(M \otimes_R M', \mathbb{Q}/\mathbb{Z}\big).$$

Proof The correspondence between

$$f : M \longrightarrow \mathsf{Ab}(M', \mathbb{Q}/\mathbb{Z}) \quad \text{and} \quad g : M \otimes_R M' \longrightarrow \mathbb{Q}/\mathbb{Z}$$

is simply

$$f(m)(m') = g(m \otimes m').$$

Observe that g is defined without ambiguity. Indeed, if $r \in R$ and $m, m' \in M$,

$$g(m \otimes rm') = f(m)(rm') = \big(r f(m)\big)(m') = f(rm)(m') = g(rm \otimes m')$$

by definition of the R-module structure of $\mathsf{Ab}(M', \mathbb{Q}/\mathbb{Z})$ and by the R-linearity of f. It is obvious that this yields an isomorphism of R-modules. $\square$

Corollary 7.10 *Let R be a ring. The object $\mathsf{Ab}(R, \mathbb{Q}/\mathbb{Z})$ is an injective cogenerator in the category Mod_R of R-modules.*

Proof The functor

$$\mathsf{Mod}_R\big(-, \mathsf{Ab}(R, \mathbb{Q}/\mathbb{Z})\big) : \mathsf{Mod}_R \longrightarrow \mathsf{Mod}_R$$

is isomorphic to the functor

$$\mathsf{Ab}(-, \mathbb{Q}/\mathbb{Z}) : \mathsf{Mod}_R \longrightarrow \mathsf{Mod}_R$$

as attested by Lemma 7.9, when choosing $M' = R$. By Proposition 7.8, this functor transforms injections into surjections and reflects isomorphisms. In particular this functor is faithful: if $f, g : A \longrightarrow B$ are two R-linear mappings and $k = \mathsf{Coker}\,(f, g)$, $\mathsf{Ab}(f, \mathrm{id}_{\mathbb{Q}/\mathbb{Z}}) = \mathsf{Ab}(g, \mathrm{id}_{\mathbb{Q}/\mathbb{Z}})$ implies that

$$\mathsf{Ab}\big(\mathsf{Coker}\,(f, g), \mathrm{id}_{\mathbb{Q}/\mathbb{Z}}\big) = \mathsf{Ker}\,\big(\mathsf{Ab}(f, \mathrm{id}_{\mathbb{Q}/\mathbb{Z}}), \mathsf{Ab}(g, \mathrm{id}_{\mathbb{Q}/\mathbb{Z}})\big)$$

is an isomorphism; thus $\mathsf{Coker}\,(f, g)$ is an isomorphism and $f = g$.

The "underlying set" functor $\mathsf{Mod}_R \longrightarrow \mathsf{Set}$ is faithful and preserves surjections. Thus the composite functor

$$\mathsf{Mod}_R\big(-, \mathsf{Ab}(R, \mathbb{Q}/\mathbb{Z})\big) = \mathsf{Ab}(-, \mathbb{Q}/\mathbb{Z}) : \mathsf{Mod}_R \longrightarrow \mathsf{Set}$$

transforms injections into surjections and is faithful. This proves that the object $\mathsf{Ab}(R, \mathbb{Q}/\mathbb{Z})$ is an injective cogenerator in Mod_R (see Propositions 7.4 and 7.2). □

7.2 Pure Submodules

The notion of a *flat* module M over a ring R is well-known:

For each monomorphism $s : A \rightarrowtail B$ of R-modules,

$$\mathsf{id}_M \otimes s : M \otimes_R A \longrightarrow M \otimes_R B$$

remains injective.

It is important not to confuse flatness with purity:

Definition 7.11 A monomorphism $s : A \rightarrowtail B$ of R-modules is *pure* when for each R-module M, the linear mapping

$$\mathsf{id}_M \otimes s : M \otimes_R A \longrightarrow M \otimes_R B$$

remains injective.

Clearly in Definition 7.11, we could omit the requirement "s is a monomorphism" since it is at once attested by the case $M = R$.

Proposition 7.12 *Let R be a ring. Each monomorphism $s : A \rightarrowtail B$ of R-modules admitting a retraction is pure.*

Proof If $r : B \longrightarrow A$ is such that $r \circ s = \mathsf{id}_A$, for each R-module M, $\mathsf{id}_M \otimes s$ admits the retraction $\mathsf{id}_M \otimes r$ and thus is injective. □

The following corollary explains why, in the case of vector spaces, pure monomorphisms are never considered.

Corollary 7.13 *In the case of a field K, every monomorphism of K-vector spaces is pure.*

Proof Every vector subspace $s : A \rightarrowtail B$ has a supplement $t : C \rightarrowtail B$, that is, $B \cong A \oplus C$. The projection $p_A : B \cong A \oplus C \longrightarrow A$ is a retraction of s. One concludes by Proposition 7.12. □

Clearly the situation of Proposition 7.12 is a rather trivial one. But our next result shows that being a pure monomorphism reduces to another monomorphism admitting a retraction, like in Proposition 7.12.

Theorem 7.14 *Let R be a ring and $s\colon A \rightarrowtail B$ a monomorphism of R-modules. The following conditions are equivalent:*

1. *the monomorphism s is pure;*
2. *the epimorphism*

$$\mathsf{Ab}(s,\mathbb{Q}/\mathbb{Z})\colon \mathsf{Ab}(B,\mathbb{Q}/\mathbb{Z}) \twoheadrightarrow \mathsf{Ab}(A,\mathbb{Q}/\mathbb{Z})$$

admits a section.

Proof By Proposition 7.8, $\mathsf{Ab}(s,\mathbb{Q}/\mathbb{Z})$ is indeed surjective. Let us use the shortened notation

$$\Gamma = \mathsf{Ab}(-,\mathbb{Q}/\mathbb{Z})\colon \mathsf{Mod}_R \longrightarrow \mathsf{Mod}_R$$

for the functor of the statement. We know that this contravariant functor Γ is exact and reflects isomorphisms (see Proposition 7.8).

For each R-linear mapping $f\colon M \longrightarrow M'$, we have a commutative diagram (see Lemma 7.9)

$$\begin{array}{ccc}
\mathsf{Mod}_R\big(\Gamma(M),\Gamma(M')\big) & \xrightarrow{\ \mathsf{Mod}_R\big(\Gamma(M),\Gamma(f)\big)\ } & \mathsf{Mod}_R\big(\Gamma(M),\Gamma(M)\big) \\
\theta_{\Gamma(M),M'}\Big\downarrow \cong & & \cong \Big\downarrow \theta_{\Gamma(M),M} \\
\Gamma\big(\Gamma(M)\otimes_R M'\big) & \xrightarrow[\ \Gamma(\mathrm{id}_{\Gamma(M)}\otimes f)\ ]{} & \Gamma\big(\Gamma(M)\otimes_R M\big)
\end{array}$$

because, for all $g\colon \Gamma(M) \longrightarrow \Gamma(M')$, $h \in \Gamma(M)$ and $m \in M$,

$$\Big(\theta_{\Gamma(M),M} \circ \mathsf{Mod}_R\big(\Gamma(M),\Gamma(f)\big)\Big)(g)(h \otimes m)$$

$$\begin{aligned}
&= \theta_{\Gamma(M),M}\big(\Gamma(f)\circ g\big)(h \otimes m) \\
&= \big(\Gamma(f)\circ g\big)(h)(m) \\
&= \big(g(h)\circ f\big)(m) \\
&= g(h)\big(f(m)\big) \\
&= \theta_{\Gamma(M),M'}(g)\big(h \otimes f(m)\big) \\
&= \big(\Gamma(\mathrm{id}_{\Gamma(M)}\otimes f)\circ \theta_{\Gamma(M),M}\big)(g)(h \otimes m).
\end{aligned}$$

$(1 \Rightarrow 2)$ because, applying Proposition 7.8,

$$\begin{aligned}
f \text{ pure} &\Rightarrow \mathrm{id}_{\Gamma(M)}\otimes f \text{ injective} \\
&\Rightarrow \Gamma\big(\mathrm{id}_{\Gamma(M)}\otimes f\big) \text{ surjective} \\
&\Rightarrow \mathsf{Mod}_R\big(\Gamma(M),\Gamma(f)\big) \text{ surjective} \\
&\Rightarrow \exists g\colon \Gamma(M) \longrightarrow \Gamma(M')\ \ \Gamma(f)\circ g = \mathrm{id}_{\Gamma(M)} \\
&\Rightarrow \Gamma(f) \text{ has a section.}
\end{aligned}$$

$(2 \Rightarrow 1)$. When $\Gamma(f)$ admits a section g and N is an R-module, the morphism

$$\mathsf{Mod}_R\left(\mathsf{id}_N, \Gamma(f)\right): \mathsf{Mod}_R\left(N, \Gamma(M')\right)\xrightarrow{\hspace{2cm}}\mathsf{Mod}_R\left(N, \Gamma(M)\right)$$

admits the section $\mathsf{Mod}_R(\mathsf{id}_N, g)$. By Lemma 7.9, the morphism

$$\Gamma(\mathsf{id}_N \otimes f): \Gamma(N \otimes_R M')\xrightarrow{\hspace{2cm}}\Gamma(N \otimes_R M)$$

has a section, thus is surjective. We thus have a short exact sequence

$$0\xrightarrow{\hspace{1.5cm}}\mathsf{Ker}\,(\mathsf{id}_N \otimes f)\xrightarrow{\hspace{0.5cm}k\hspace{0.5cm}}N \otimes_R M'\xrightarrow{\mathsf{id}_N \otimes f}N \otimes_R M\xrightarrow{\hspace{1.5cm}}0$$

with $k = \ker(\mathsf{id}_N \otimes f)$; by Proposition 7.8

$$\Gamma\big(\mathsf{Ker}\,(\mathsf{id}_N \otimes f)\big) \cong \mathsf{Coker}\,\big(\Gamma(\mathsf{id}_N \otimes f)\big) \cong 0.$$

Therefore $\mathsf{Ker}\,(\mathsf{id}_N \otimes f) \cong 0$ because Γ reflects isomorphisms (see Proposition 7.8) and thus $\mathsf{id}_N \otimes f$ is a monomorphism. □

7.3 What is the Descent Problem for Rings?

The following result concerning modules over a ring is classical:

Proposition 7.15 *Let* $\sigma: R\longrightarrow S$ *be a ring homomorphism.*

1. *There exists a functor*

$$\mathsf{U}_\sigma: \mathsf{Mod}_S\xrightarrow{\hspace{2cm}}\mathsf{Mod}_R, \quad B \mapsto B$$

of restriction of the scalars, where

$$r \cdot b = \sigma(r) \cdot b, \quad r \in R, \ b \in B.$$

2. *There exists a functor*

$$S \otimes_R -: \mathsf{Mod}_R\xrightarrow{\hspace{2cm}}\mathsf{Mod}_S, \quad A \mapsto S \otimes_R A$$

of extension of the scalars, where

$$s'(s \otimes a) = (s's) \otimes a, \quad s, s' \in S, \ a \in A.$$

3. *The functor* $S \otimes_R -$ *is left adjoint to the functor* U_σ.
4. *The functor* U_σ *is monadic.*

Proof It is obvious that U_σ and $S \otimes_R -$ are functors. To prove the adjunction, consider objects $A \in \mathsf{Mod}_R$ and $B \in \mathsf{Mod}_S$. The morphisms

$$\varepsilon_A: A\longrightarrow S \otimes_R A, \quad a \mapsto 1 \otimes a$$
$$\eta_B: S \otimes_R B\longrightarrow B, \quad s \otimes b \mapsto sb$$

define two natural transformations

$$\varepsilon: \mathsf{id}_{\mathsf{Mod}_R} \Rightarrow \mathsf{U}_\sigma(S \otimes_R -), \quad \eta: S \otimes \mathsf{U}_\sigma(-) \Rightarrow \mathsf{id}_{\mathsf{Mod}_S}.$$

One immediately sees that they satisfy the conditions of Proposition 5.4.

The functor U_σ reflects isomorphisms, because in both categories the isomorphisms are the bijective morphisms. Moreover in both categories, coequalizers exist and are computed as in the category of abelian groups; thus U_σ preserves them. We can conclude by applying the Beck criterion (see Theorem 5.17).   □

Theorem 7.15 is in fact a special case of a more general theorem which asserts that every algebraic functor is monadic.

Thus in Proposition 7.15, the functor U_σ is always monadic, without any assumption on the ring homomorphism σ. The classical problem of effective descent can then elegantly be described as:

Definition 7.16 A morphism $\sigma: R \longrightarrow S$ of rings is of *effective descent* when the functor

$$S \otimes_R -: \mathsf{Mod}_R \longrightarrow \mathsf{Mod}_S, \quad A \mapsto S \otimes_R A$$

is comonadic.

It is useful to describe explicitly the situation of Definition 7.16. The comonad $\mathbb{S} = (S, \varphi, \eta)$ associated with the adjunction of Proposition 7.15 is given by, for each $B \in \mathsf{Mod}_S$:

$$S: \mathsf{Mod}_S \longrightarrow \mathsf{Mod}_S, \quad B \mapsto S \otimes_R B,$$
$$\varphi_B: S \otimes_R B \longrightarrow B, \quad s \otimes b \mapsto sb,$$
$$\zeta_B: S \otimes_R B \longrightarrow S \otimes_R S \otimes_R B, \quad s \otimes b \mapsto s \otimes 1 \otimes b.$$

The category of $\mathbb{S}$-coalgebras is denoted by Desc_σ and is also called the "category of descent data". An object in it is a pair (B, ξ) where

$$B \in \mathsf{Mod}_S, \quad \xi: B \longrightarrow S \otimes_R B$$

satisfy the conditions dual to those of Definition 5.9, that is, the commutativity of the diagrams

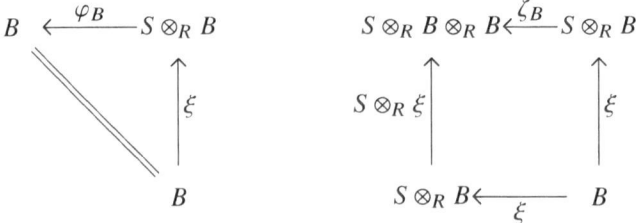

A morphism $f: (B, \xi) \longrightarrow (B', \xi')$ is then a morphism $f: B \longrightarrow B'$ in Mod_S such that the following square commutes:

$$S \otimes_R B' \xleftarrow{\;S \otimes_R f\;} S \otimes_R B$$

$$\xi' \Big\uparrow \qquad\qquad \Big\uparrow \xi$$

$$B' \xleftarrow{\quad f \quad} B$$

The functor $S \otimes_R -$ is thus comonadic when (the dual situation of Definition 5.13) the functor J obtained by duality from Proposition 5.12 is an equivalence of categories. This functor has the form

$$J: \mathsf{Mod}_R \longrightarrow \mathsf{Desc}_\sigma, \quad A \mapsto (S \otimes_R A, \mathsf{id}_S \otimes \varepsilon_A), \quad f \mapsto \mathsf{id}_S \otimes_R f$$

where ε is defined in Proposition 7.15. More explicitly

$$\mathsf{id}_S \otimes \varepsilon_A: S \otimes_R A \longrightarrow S \otimes_R S \otimes_R A, \quad s \otimes a \mapsto s \otimes 1 \otimes a.$$

Let us mention an apparently weaker notion:

Definition 7.17 A morphism $\sigma: R \longrightarrow S$ of rings is called a *descent morphism* when the comparison functor

$$J: \mathsf{Mod}_R \longrightarrow \mathsf{Desc}_\sigma, \quad A \mapsto (S \otimes_R A, \mathsf{id}_S \otimes \varepsilon_A)$$

is full and faithful.

However, as we shall see in Theorem 7.20, in the case of commutative rings with unit, "descent" and "effective descent" are equivalent.

To conclude this short introduction to the descent problem for rings, let us mention that in the literature, this problem can appear in rather different forms. That is, one defines first a notion of descent data which, often, is not that of being a coalgebra as above; one defines next a corresponding comparison functor J and the morphism of rings is declared of effective descent when this functor J is an equivalence of categories. But the resulting category of descent data turns always turns out to be equivalent to the category of $\mathbb{S}$-coalgebras and thus, the problem of effective descent remains the same.

7.4 The Descent Theorem for Rings

This section characterizes the morphisms of rings which are of effective descent: they are exactly the pure ones.

Definition 7.18 A *pure extension* of rings is a ring homomorphism $\sigma: R \rightarrowtail S$ which is a pure monomorphism of R-modules.

Let us recall that the ring S, like every S-module, can indeed be seen as an R-module $U_\sigma(S)$ (see Proposition 7.15).

Proposition 7.19 *Every extension of fields is pure.*

Proof By Corollary 7.13.   □

Theorem 7.20 *Let $\sigma: R \longrightarrow S$ be a morphism of rings. The following conditions are equivalent:*

1. *σ is a morphism of effective descent;*
2. *σ is a morphism of descent;*
3. *σ is a pure extension of rings.*

Proof $(1 \Rightarrow 2)$ is trivial. Let us prove $(2 \Rightarrow 3)$. If σ is a descent morphism, the comparison functor J of Definition 7.17 is full and faithful. But each faithful functor F reflects monomorphisms: indeed if $F(f)$ is a monomorphism

$$f \circ u = f \circ v \Rightarrow F(f) \circ F(u) = F(f) \circ F(v) \Rightarrow F(u) = F(v) \Rightarrow u = v.$$

We must prove that for each R-module A, the morphism

$$\varepsilon_A = \sigma \otimes \mathrm{id}_A: A \cong R \otimes_R A \longrightarrow S \otimes_R A, \quad a \mapsto 1 \otimes a$$

is a monomorphism of R-modules. Since the functor J is faithful, it thus suffices to prove that

$$J(\varepsilon_A) = \mathrm{id}_S \otimes \varepsilon_A: \left(S \otimes_R A, \mathrm{id}_S \otimes \varepsilon_A\right) \longrightarrow \left(S \otimes_R S \otimes_R A, \mathrm{id}_S \otimes \varepsilon_{S \otimes_R A}\right)$$

is a monomorphism in Desc_σ. Trivially, it suffices to prove that the morphism

$$\mathrm{id}_S \otimes \alpha_A: S \otimes_R A \longrightarrow S \otimes_R S \otimes_R A, \quad s \otimes a \mapsto s \otimes 1 \otimes a$$

is injective. Indeed this mapping is injective, since it admits the restriction

$$\eta_S \otimes \mathrm{id}_A: S \otimes_R S \otimes_R A \longrightarrow S \otimes_R A, \quad s \otimes s' \otimes a \mapsto ss' \otimes a.$$

$(3 \Rightarrow 1)$. The assumption is now that for each R-module A, the morphism

$$\varepsilon_A: A \cong R \otimes_R A \longrightarrow S \otimes_R A, \quad a \mapsto 1 \otimes a$$

is injective. We shall use the Beck criterion (or more precisely, its dual; see Theorem 5.17) to prove that the functor

$$S \otimes_R -: \mathsf{Mod}_R \longrightarrow \mathsf{Mod}_S$$

is comonadic. We know already that $S \otimes_R -$ admits U_σ as right adjoint functor (see Proposition 7.15).

First, let us prove that $S \otimes_R -$ reflects isomorphisms. The isomorphisms of modules are the bijective morphisms, the monomorphisms are the injective ones and the epimorphisms, the surjective ones. Thus it suffices to prove that $S \otimes_R -$ reflects both monomorphisms and epimorphisms.

The functor $S \otimes_R -$ reflects the zero morphisms:

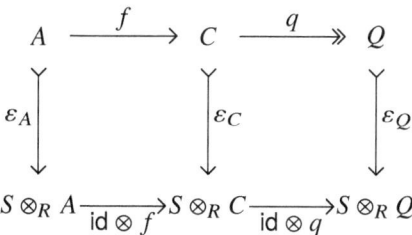

if $\mathsf{id} \otimes f = 0$, then $\varepsilon_C \circ f = 0$, thus $f = 0$ because ε_C is a monomorphism.

If $\mathsf{id} \otimes f$ is an epimorphism, let us consider $q = \mathsf{Coker}\, f$. The equality $q \circ f = 0$ implies $(\mathsf{id} \otimes q) \circ (\mathsf{id} \otimes f) = 0$. Thus $\mathsf{id} \otimes q = 0$ because $\mathsf{id} \otimes f$ is an epimorphism; this forces $\varepsilon_Q \circ q = 0$, thus $q = 0$ since ε_Q is a monomorphism. So $\mathsf{Coker}\, f = 0$ and f is an epimorphism.

Since ε_A is a monomorphism by assumption, if also $\mathsf{id} \otimes f$ is a monomorphism, the composite $(\mathsf{id} \otimes f) \circ \varepsilon_A = \varepsilon_C \circ f$ is a monomorphism, thus f is a monomorphism.

To prove the condition on split coequalizers, we must consider two morphisms (u, v) in Mod_R such that the pair $(\mathsf{id}_S \otimes u, \mathsf{id}_S \otimes v)$ admits a split coequalizer l in Mod_S.

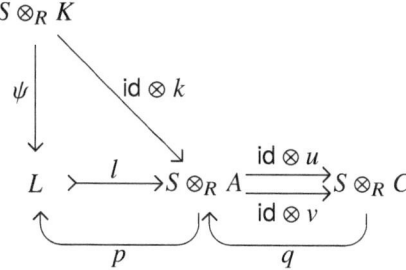

Thus, there exist morphisms l, p, q in Mod_S such that

$$(\mathsf{id} \otimes u) \circ l = (\mathsf{id} \otimes v) \circ l, \quad p \circ l = \mathsf{id}, \quad q \circ (\mathsf{id} \otimes u) = \mathsf{id}, \quad q \circ (\mathsf{id} \otimes v) = l \circ p.$$

In the category Mod_R, consider the equalizer $k = \mathsf{Ker}\,(u, v)$. Consider further in Mod_S the factorization ψ of $\mathsf{id} \otimes k$ through the equalizer $l = \mathsf{Ker}\,(\mathsf{id} \otimes u, \mathsf{id} \otimes v)$. We must prove that ψ is an isomorphism.

For this, consider the following diagram constituted of categories and functors:

$$\begin{array}{ccc}
\mathsf{Mod}_R & \xrightarrow{\;S \otimes_R -\;} & \mathsf{Mod}_S \\
\mathsf{Ab}(-,\mathbb{Q}/\mathbb{Z}) \downarrow & \cong & \downarrow \mathsf{Ab}(-,\mathbb{Q}/\mathbb{Z}) \\
\mathsf{Mod}_R & \xrightarrow{\quad\quad} & \mathsf{Ab} \\
& \mathsf{Mod}_R(S, -) &
\end{array}$$

By Lemma 7.9, this diagram is commutative (up to an isomorphism). By Proposition 7.8, both contravariant functors $\mathsf{Ab}(-,\mathbb{Q}/\mathbb{Z})$ are exact and reflect isomorphisms.

By Theorem 7.14 and Lemma 7.9, we have the following situation, for each R-module $A \cong R \otimes_R A$

$$\begin{array}{ccc}
\mathsf{Ab}(A,\mathbb{Q}/\mathbb{Z}) & \underset{\rho_A}{\overset{\tau_A}{\rightleftarrows}} & \mathsf{Ab}(S \otimes_R A, \mathbb{Q}/\mathbb{Z}) \\
\cong \downarrow & & \downarrow \cong \\
\mathsf{Mod}_R(A, \mathsf{Ab}(R,\mathbb{Q}/\mathbb{Z})) & \underset{\mathsf{Mod}_R(\mathsf{id}, \mathsf{Ab}(s,\mathsf{id}))}{\overset{\mathsf{Mod}_R(\mathsf{id}, g)}{\rightleftarrows}} & \mathsf{Mod}_R(A, \mathsf{Ab}(S,\mathbb{Q}/\mathbb{Z}))
\end{array}$$

where the vertical morphisms are isomorphisms and $\mathsf{Ab}(s,\mathsf{id}) \circ g = \mathsf{id}$ (see Theorem 7.14). This situation yields the composites ρ_A and τ_A, which are natural in A and such that $\rho_A \circ \tau_A = \mathsf{id}$.

In the category of R-modules, we obtain the following situation.

$$\begin{array}{ccccc}
\mathsf{Ab}(K,\mathbb{Q}/\mathbb{Z}) & \xleftarrow{\mathsf{Ab}(k,\mathsf{id})} & \mathsf{Ab}(A,\mathbb{Q}/\mathbb{Z}) & \underset{\mathsf{Ab}(v,\mathsf{id})}{\overset{\mathsf{Ab}(u,\mathsf{id})}{\leftleftarrows}} & \mathsf{Ab}(C,\mathbb{Q}/\mathbb{Z}) \\
\tau \uparrow\downarrow \rho & & \tau_A \uparrow\downarrow \rho_A & & \tau_C \uparrow\downarrow \rho_C \\
\mathsf{Ab}(L,\mathbb{Q}/\mathbb{Z}) & \xleftarrow{\mathsf{Ab}(l,\mathsf{id})} & \mathsf{Ab}(S \otimes_R A, \mathbb{Q}/\mathbb{Z}) & \underset{\mathsf{Ab}(\mathsf{id}\otimes v,\mathsf{id})}{\overset{\mathsf{Ab}(\mathsf{id}\otimes u,\mathsf{id})}{\leftleftarrows}} & \mathsf{Ab}(S \otimes_R C, \mathbb{Q}/\mathbb{Z})
\end{array}$$

with $\overline{p}$, $\overline{q}$ at top and $\mathsf{Ab}(p,\mathsf{id})$, $\mathsf{Ab}(q,\mathsf{id})$ at bottom.

The bottom line is a split coequalizer. The top line is a coequalizer, because the contravariant functor $\mathsf{Ab}(-,\mathbb{Q}/\mathbb{Z})$ is exact (see Proposition 7.8). The commutativity of the right hand part implies the existence of the factorizations τ and ρ through these coequalizers, so that the whole diagram remains commutative, with $\rho \circ \tau = \mathsf{id}$. It suffices now to define

$$\overline{p} = \rho_A \circ \mathsf{Ab}(p,\mathsf{id}) \circ \tau, \quad \overline{q} = \rho_C \circ \mathsf{Ab}(q,\mathsf{id}) \circ \tau_A$$

to present the first line as a split coequalizer. Indeed,

$$\mathsf{Ab}(k, \mathsf{id}) \circ \overline{p} = \mathsf{Ab}(k, \mathsf{id}) \circ \rho_A \circ \mathsf{Ab}(p, \mathsf{id}) \circ \tau$$
$$= \rho \circ \mathsf{Ab}(l, \mathsf{id}) \circ \mathsf{Ab}(p, \mathsf{id}) \circ \tau$$
$$= \rho \circ \mathsf{id} \circ \tau$$
$$= \mathsf{id},$$
$$\mathsf{Ab}(u, \mathsf{id}) \circ \overline{q} = \mathsf{Ab}(u, \mathsf{id}) \circ \rho_C \circ \mathsf{Ab}(q, \mathsf{id}) \circ \tau_A$$
$$= \rho_A \circ \mathsf{Ab}(\mathsf{id} \otimes u, \mathsf{id}) \circ \mathsf{Ab}(q, \mathsf{id}) \circ \tau_A$$
$$= \rho_A \circ \mathsf{id} \circ \tau_A$$
$$= \mathsf{id},$$
$$\mathsf{Ab}(v, \mathsf{id}) \circ \overline{q} = \mathsf{Ab}(v, \mathsf{id}) \circ \rho_C \circ \mathsf{Ab}(q, \mathsf{id}) \circ \tau_A$$
$$= \rho_A \circ \mathsf{Ab}(\mathsf{id} \otimes v, \mathsf{id}) \circ \mathsf{Ab}(q, \mathsf{id}) \circ \tau_A$$
$$= \rho_A \circ \mathsf{Ab}(p, \mathsf{id}) \circ \mathsf{Ab}(l, \mathsf{id}) \circ \tau_A$$
$$= \rho_A \circ \mathsf{Ab}(p, \mathsf{id}) \circ \tau \circ \mathsf{Ab}(k, \mathsf{id})$$
$$= \overline{p} \circ \mathsf{Ab}(k, \mathsf{id}).$$

The contravariant functor $\mathsf{Ab}(-, \mathbb{Q}/\mathbb{Z})$ is exact (see Proposition 7.8), thus transforms $l = \mathsf{Ker}\,(\mathsf{id} \otimes u, \mathsf{id} \otimes v)$ into the coequalizer

$$\mathsf{Coker}\,\big(\mathsf{Ab}(\mathsf{id}_S \otimes u, \mathsf{id}_{\mathbb{Q}/\mathbb{Z}}), \mathsf{Ab}(\mathsf{id}_S \otimes v, \mathsf{id}_{\mathbb{Q}/\mathbb{Z}})\big).$$

Moreover, by commutativity of the diagram of functors and categories above,

$$\mathsf{Ab}\big(\mathsf{id}_S \otimes k, \mathsf{id}_{\mathbb{Q}/\mathbb{Z}}\big) = \mathsf{Mod}_R\big(\mathsf{id}_S, \mathsf{Ab}(k, \mathsf{id}_{\mathbb{Q}/\mathbb{Z}})\big)$$

is a split coequalizer, because in Mod_R, the coequalizer

$$\mathsf{Ab}(k, \mathsf{id}_{\mathbb{Q}/\mathbb{Z}}) = \mathsf{Coker}\,\big(\mathsf{Ab}(u, \mathsf{id}_{\mathbb{Q}/\mathbb{Z}}), \mathsf{Ab}(v, \mathsf{id}_{\mathbb{Q}/\mathbb{Z}})\big)$$

is split. By uniqueness of a coequalizer in the category Ab of abelian groups, the factorization $\mathsf{Ab}(\psi, \mathsf{id}_{\mathbb{Q}/\mathbb{Z}})$ is an isomorphism in Ab. But the functor $\mathsf{Ab}(-, \mathbb{Q}/\mathbb{Z})$ reflects isomorphisms (see Proposition 7.8); thus ψ is an isomorphism in Mod_S. $\quad\square$

7.5 Algebras Over a Ring

This section recalls some classical results of the theory of algebras over a ring.

Definition 7.21 Let R be a ring. An R-algebra is a sextuple $(A, +, \times, \cdot, 0, 1)$ where

1. $(A, +, \times, 0, 1)$ is a ring, commutative, with unit;
2. $(A, +, \cdot)$ is an R-module;
3. $\forall r \in R \;\; \forall a, a' \in A \;\; r \cdot (a \times a') = (r \cdot a) \times a'$.

In general, we simply write $r \cdot a = ra$ and $a \times a' = aa'$. An ideal $I \lhd A$ of an R-algebra is an ideal I of the ring $(A, +, \times)$; in particular, I is an R-submodule of A:

$$\forall r \in R \ \ \forall i \in I \ \ \ r \cdot i = r \cdot (1 \times i) = (r \cdot 1) \times i \in I.$$

The quotient A/I is still an R-algebra.

Observe that trivially:

Lemma 7.22 *The category of rings is also the category of $\mathbb{Z}$-algebras.* $\square$

Proposition 7.23 *The category Alg_R of R-algebras over the ring R is isomorphic to the category R/Ring of rings under R, that is:*

- *objects: the pairs (A, α) where A is a ring and $\alpha \colon R \longrightarrow A$ is a morphism of rings;*
- *morphisms: $f \colon (A, \alpha) \longrightarrow (B, \beta)$ is a morphism $f \colon A \longrightarrow B$ of rings such that $f \circ \alpha = \beta$.*

More generally, when S is an R-algebra, the category of S-algebras is isomorphic to the category of R-algebras under S.

Proof Each R-algebra A yields a morphism of rings

$$i_A \colon R \longrightarrow A, \ \ r \mapsto r1.$$

Conversely, given such a homomorphism of rings, it suffices to define

$$r \cdot a = i_A(r) \times a, \ \ r \in R, \ a \in A$$

to get an R-algebra structure on the ring A. This argument carries over as such to the more general case. $\square$

Proposition 7.24 *The category Alg_R of R-algebras over a ring R is (finitely) complete and cocomplete.*

Proof We use Proposition 3.9, in the finite case. The terminal R-algebra (the product of an empty family) is $\{0\}$: for each R-algebra A, there is indeed a unique morphism of R-algebras $A \longrightarrow \{0\}$. The product $A \times B$ of two R-algebras is their Cartesian product with the componentwise operations. This takes care of finite products.

The equalizer of two morphisms $f, g \colon A \rightrightarrows B$ of R-algebras is the subset $\{a \in A \mid f(a) = g(a)\}$ with the operations induced by those of A. The pullback of two morphisms $f \colon A \longrightarrow B$ and $g \colon C \longrightarrow B$ is the set $\{(a, c) \mid f(a) = g(c)\}$ with the componentwise operations.

The initial R-algebra is R itself: for each R-algebra A there is a morphism of R-algebras

$$i_A \colon R \longrightarrow A, \ \ r \mapsto r \cdot 1;$$

and this morphism is unique because necessarily, given $j \colon R \longrightarrow A$, $j(1) = 1$, thus $j(r) = j(r1) = r \cdot 1$.

The coproduct of two R-algebras A, B is their tensor product $A \otimes_R B$ with the multiplication

$$\left(\sum_i a_i \otimes b_i\right) \cdot \left(\sum_j a'_j \otimes b'_j\right) = \sum_{i,j} a_i a'_j \otimes b_i b'_j$$

and with the canonical morphisms

$$A \longrightarrow A \otimes_R B, \ a \mapsto a \otimes 1, \quad B \longrightarrow A \otimes_R B, \ b \mapsto 1 \otimes b.$$

Indeed if $f: A \longrightarrow C$ and $g: B \longrightarrow C$ are morphisms of R-algebras, the unique factorization through the tensor product is given by

$$h: A \otimes_R B \longrightarrow C, \quad \sum_i a_i \otimes b_i \mapsto \sum_i a_i b_i.$$

In the same way, the pushout of two morphisms of R-algebras is their tensor product over their common domain:

$$
\begin{array}{ccc}
A & \xrightarrow{\ f\ } & B \\
\downarrow{\scriptstyle g} & & \downarrow{\scriptstyle \mathrm{id}_B \otimes 1} \\
C & \xrightarrow[1 \otimes \mathrm{id}_C]{} & B \otimes_A C
\end{array}
$$

Finally the coequalizer of two morphisms $f, g: A \rightrightarrows B$ of R-algebras is the quotient of B by the ideal generated by the elements $f(a) - g(a)$, for each $a \in A$.

This proves the existence of finite limits and finite colimits. Arbitrary limits and colimits exist as well, as in every algebraic category, but we shall not need them in this book. $\square$

7.6 The Descent Theory of Algebras

The descent theory of algebras is just a consequence of the descent theory of modules, as studied in Sections 7.3 and 7.4.

Proposition 7.25 *Let $\sigma: R \longrightarrow S$ be a morphism of rings.*

1. *There exists a functor*

$$U_\sigma: \mathsf{Alg}_S \longrightarrow \mathsf{Alg}_R, \quad B \mapsto B$$

of restriction of the scalars, where

$$r \cdot b = \sigma(r) \cdot b, \quad r \in R, \ b \in B.$$

2. *There exists a functor*

$$S \otimes_R - : \mathsf{Alg}_R \longrightarrow \mathsf{Alg}_S, \quad A \mapsto S \otimes_R A$$

of extension of the scalars, *where*

$$\left(\sum_i s_i \otimes a_i \right) \cdot \left(\sum_j s'_j \otimes a'_j \right) = \sum_{i,j} (s_i s'_j) \otimes (a_i a'_j),$$

$$s' \left(\sum_i s_i \otimes a_i \right) = \sum_i (s' s_i) \otimes a_i$$

with $s_i, s'_j, s' \in S$, $a_i, a'_j \in A$.
3. *The functor* $S \otimes_R -$ *is left adjoint to the functor* U_σ.
4. *the functor* U_σ *is monadic.*

Proof It is obvious that U_σ and $S \otimes_R -$ are functors. To prove the adjunction, consider objects $A \in \mathsf{Alg}_R$ and $B \in \mathsf{Alg}_S$. The morphisms

$$\varepsilon_A : A \longrightarrow S \otimes_R A, \quad a \mapsto 1 \otimes a$$
$$\eta_B : S \otimes_R B \longrightarrow B, \quad s \otimes b \mapsto sb$$

define two natural transformations

$$\varepsilon : \mathsf{id}_{\mathsf{Alg}_R} \Rightarrow \mathsf{U}_\sigma (S \otimes_R -), \quad \eta : S \otimes \mathsf{U}_\sigma (-) \Rightarrow \mathsf{id}_{\mathsf{Alg}_S}.$$

We see immediately that these natural transformations satisfy the conditions of Proposition 5.4.

The functor U_σ reflects isomorphisms, because in both categories they are the bijective morphisms. Moreover both categories admit coequalizers computed as in the category of rings, thus the functor U_σ preserves these coequalizers. The conclusion follows now from Beck's criterion (see Theorem 5.17).   □

Proposition 7.26 *Let* $\sigma : R \longrightarrow S$ *be a morphism of rings. The following conditions are equivalent:*

1. $\sigma : R \longrightarrow S$ *is a pure extension of rings;*
2. *for each R-algebra A, the morphism*

$$\varepsilon_A = \sigma \otimes \mathsf{id}_A : A \cong R \otimes_R A \longrightarrow S \otimes_R A$$

is injective.

Proof $(1 \Rightarrow 2)$ because each R-algebra A is in particular an R-module. Conversely, observe that each R-module N is a submodule of the R-algebra $A = R \oplus N$ where the multiplication of A is defined by

$$(r, n)(r', n') = (rr', rn' + r'n).$$

The module A, provided with this multiplication, is indeed an R-algebra:

$$
\begin{aligned}
(r,n)\big((r',n')(r'',n'')\big) &= (r,n)(r'r'',r'n''+r''n') \\
&= (rr'r'',rr'n''+rr''n'+r'r''n) \\
&= (rr',rn'+r'n)(r'',n'') \\
&= \big((r,n)(r',n')\big)(r'',n''); \\
(r,n)(r',n') &= (rr',rn'+r'n) \\
&= (r',n')(r,n); \\
(r,n)(1,0) &= (r,n); \\
(r,n)\big((r',n')+(r'',n'')\big) &= (r,n)(r'+r'',n'+n'') \\
&= (rr'+rr'',rn'+rn''+r'n+r''n) \\
&= (rr',rn'+r'n)+(rr'',rn''+r''n) \\
&= (r,n)(r',n')+(r,n)(r'',n''); \\
r\big((r',n')(r'',n'')\big) &= r(r'r'',r'n''+r''n') \\
&= (rr'r'',rr'n''+rr''n') \\
&= (rr',rn')(r'',n'') \\
&= \big(r(r',n')\big)(r'',n'').
\end{aligned}
$$

Moreover, the morphism

$$
s_N : N \longrightarrow R \oplus N, \quad n \mapsto (0,n)
$$

is an injection of R-modules. Consider again

$$
\varepsilon_N : N \longrightarrow S \otimes_R N, \quad n \mapsto 1 \otimes n.
$$

In the diagram

$$
\begin{array}{ccc}
N & \xrightarrow{\;s_N\;} & R \oplus N \\
\downarrow{\scriptstyle \varepsilon_N} & & \downarrow{\scriptstyle \varepsilon_{R \otimes N}} \\
S \otimes_R N & \xrightarrow[\mathrm{id} \otimes s_N]{} & S \otimes_R (R \oplus N)
\end{array}
$$

the morphism $\varepsilon_{R \oplus N}$ is injective, because $A = R \oplus N$ is an R-algebra; and the morphism s_N is injective as well. Thus ε_N is injective. $\square$

Theorem 7.27 *Let $\sigma : R \longrightarrow S$ be a ring homomorphism. The following conditions are equivalent:*

1. *$\sigma : R \longrightarrow S$ is a pure extension of rings;*
2. *the functor $S \otimes_R - : \mathsf{Alg}_R \longrightarrow \mathsf{Alg}_S$ is comonadic.*

Proof $(1 \Rightarrow 2)$. By Theorem 7.20, we know that the functor

$$S \otimes_R - : \mathsf{Mod}_R \longrightarrow \mathsf{Mod}_S$$

is comonadic. But each algebra is in particular a module and each morphism of algebras is in particular a morphism of modules. Moreover a morphism of algebras is an isomorphism precisely when it is bijective, that is, when it is an isomorphism of modules. Next a diagram of algebras is an equalizer precisely when it is an equalizer in the category of modules, or even in the category of sets. This proves that when the functor $S \otimes_R -$, in the case of modules, satisfies the conditions of the dual of Beck's criterion (see Theorem 5.17), the functor $S \otimes_R -$, in the case of algebras, satisfies them as well.

$(2 \Rightarrow 1)$. Let A be an R-algebra. Consider the morphism

$$\varepsilon_A : A \longrightarrow S \otimes_R A, \quad a \mapsto 1 \otimes a.$$

The image of ε_A by the functor $S \otimes_R -$

$$\mathsf{id}_S \otimes \varepsilon_A : S \otimes_R A \longrightarrow S \otimes_R S \otimes_R A, \quad s \otimes a \mapsto s \otimes 1 \otimes a$$

is injective because it admits the R-linear restriction

$$S \otimes_R S \otimes_R A \longrightarrow S \otimes_R A, \quad s \otimes s' \otimes a \mapsto ss' \otimes a.$$

When $S \otimes_R -$ is comonadic, this functor reflects monomorphisms (see the dual of Proposition 5.10), thus ε_A is a monomorphism, for each R-algebra A. We conclude by Proposition 7.26. $\square$

Chapter 8
The Pierce Spectrum of a Ring

Convention. *In this chapter, all rings and algebras are commutative with a unit.*

Abstract A spectrum of a ring is a topological space associated with the ring which allows, up to an isomorphism, a representation of the ring as a ring of continuous functions on that spectrum. Many different spectra of a ring can be defined, each of them allowing a different representation of the ring, with particular properties. We are interested here in the *Pierce spectrum* of the ring. The idempotent elements of a commutative ring with unit constitute a Boolean algebra and the profinite space, corresponding to that Boolean algebra via the Stone duality, is the Pierce spectrum of the ring. The properties of this profinite Pierce spectrum, in the case of algebras over a ring, will play an essential role in developing the Galois theory of rings. When the ring is a field K, its spectrum is just a singleton and thus is never considered. Nevertheless, the spectra of the algebras over the field K were implicitly present in the Galois theory of fields.

8.1 The Boolean Algebra of Idempotent Elements

Let us recall that:

Definition 8.1 An element $e \in R$ of a ring is idempotent when $e^2 = e$.

Proposition 8.2 *The idempotent elements of a ring R, with the operations*

$$e \wedge e' = ee', \quad e \vee e' = e + e' - ee', \quad \complement e = 1 - e,$$

constitute a Boolean algebra $\mathsf{Idemp}(R)$.

© The Author(s), under exclusive license to Springer Nature Switzerland AG 2024
F. Borceux, *Galois Theories of Fields and Rings*, Coimbra Mathematical Texts 2,
https://doi.org/10.1007/978-3-031-58460-2_9

Proof If e, e' are idempotent,

$$(ee')^2 = e^2 e'^2 = ee',$$
$$(e + e' - ee')^2 = e^2 + e'^2 + e^2 e'^2 + 2ee' - 2e^2 e' - 2ee'^2$$
$$= e + e' + ee' + 2ee' - 2ee' - 2ee'$$
$$= e + e' - ee'$$

thus ee' and $e + e' - ee'$ are still idempotent.

Next observe that

$$e \wedge e' = e \Leftrightarrow ee' = e \Leftrightarrow e' = e + e' - ee' \Leftrightarrow e' = e \vee e'.$$

We consider therefore the relation $\leq$:

$$e \leq e' \text{ iff } e \wedge e' = e \text{ iff } e \vee e' = e'.$$

This defines a partial order on the set of idempotent elements:

$$ee = e \Rightarrow e \leq e;$$
$$e \leq e' \leq e'' \Rightarrow ee' = e, \ e'e'' = e'$$
$$\Rightarrow e = ee' = ee'e'' = ee''$$
$$\Rightarrow e \leq e''$$
$$e \leq e', \ e' \leq e \Rightarrow e = ee', \ e' = e'e$$
$$\Rightarrow e = e'.$$

For this partial order, $\wedge$ and $\vee$ are thus the *infimum* and *supremum* operations. Clearly, $0e = 0$ implies that 0 is the bottom element, while $e1 = e$ proves that 1 is the top element.

The distributivity law

$$e \wedge (e' \vee e'') = (e \wedge e') \vee (e \wedge e'')$$

means

$$e(e' + e'' - e'e'') = ee' + ee'' - e^2 e'e''$$

and it holds since $e = e^2$. We have thus already a distributive lattice.

Let us prove now that each idempotent element e admits $1 - e$ as complement. First,

$$(1 - e)^2 = 1 - 2e + e^2 = 1 - 2e + e = 1 - e$$

and thus $1 - e$ is idempotent as well. Next

$$e(1 - e) = e - e^2 = e - e = 0$$

thus $e \wedge (1 - e) = 0$, and

$$e + (1 - e) - e(1 - e) = e + 1 - e - e + e^2 = e + 1 - e - e + e = 1$$

thus $e \vee (1 - e) = 1$.   □

Definition 8.3 The Pierce spectrum $\mathsf{Pierce}(R)$ of a ring R is the spectrum of its Boolean algebra of idempotent elements (see Definition 3.23).

Proposition 8.4 *The construction of the Pierce spectrum of a ring extends as a functor*

$$\mathsf{Pierce} \colon \mathsf{Rings} \longrightarrow \mathsf{Prof}$$

from the category Rings *of commutative rings with unit to the category* Prof *of profinite topological spaces.*

Proof A morphism $f \colon R \longrightarrow R'$ of rings preserves the idempotent elements as well as the operations

$$e \wedge e' = ee', \quad e \vee e' = e + e' - ee', \quad \complement e = 1 - e.$$

It induces therefore a morphism of Boolean algebras $\mathsf{Idemp}(f)$ between the corresponding Boolean algebras of idempotent elements. Composing further with the Stone duality (see Theorem 3.29) yields the spectrum functor:

$$\mathsf{Pierce} \colon \mathsf{Rings} \xrightarrow{\;\mathsf{Idemp}\;} \mathsf{Boole} \xrightarrow{\;\cong\;} \mathsf{Prof}.$$ □

Let us conclude this section with an important property of the functor Pierce, which should be linked with the characterization of profinite spaces as in Theorem 3.18.

Proposition 8.5 *The contravariant functor*

$$\mathsf{Pierce} \colon \mathsf{Rings} \longrightarrow \mathsf{Prof}$$

transforms filtered colimits into cofiltered limits.

Proof By the Stone duality (see Theorem 3.29), it suffices to prove that the functor

$$\mathsf{Idemp} \colon \mathsf{Rings} \longrightarrow \mathsf{Boole}$$

preserves filtered colimits. In both categories, filtered colimits are constructed as in the category of sets. If $R = \mathrm{colim}_{i \in I} R_i$ is a filtered colimit of rings, an idempotent element $e = e^2 \in R$ has the form $e = [e_i]$ for some arbitrary element $e_i \in R_i$, for some index i. The equality $[e_i]^2 = [e_i]$ in R implies the existence of a morphism $f_{i,j} \colon R_i \longrightarrow R_j$ in the diagram, such that $f_{i,j}(e_i)^2 = f_{i,j}(e_i)$. Therefore $\mathsf{Idemp}(R) \subseteq \mathrm{colim}_{i \in I} \mathsf{Idemp}(R_i)$. The other inclusion is obvious.   □

8.2 The Regular Ideals

Definition 8.6 An ideal $I \lhd R$ of a ring is *regular* when it is generated by its idempotent elements. As usual, a *maximal* regular ideal is a maximal element in the partially ordered set of proper regular ideals.

Lemma 8.7 *An ideal $I \lhd R$ of a ring is regular when each element $i \in I$ can be written in the form*

$$i = r_1 e_1 + \cdots + r_n e_n$$

with $r_i \in R$, $e_i \in I$, $e_i^2 = e_i$.

Proof The condition in the statement means indeed that I, as an R-module, is generated by its idempotent elements.   □

Proposition 8.8 *Let $I \lhd R$ be an ideal in a ring. The following conditions are equivalent:*

1. *the ideal $I \lhd R$ is regular;*
2. $\forall i \in I$ $\exists e \in I$ $e = e^2$, $i = ie$;
3. $\forall i_1, \ldots, i_n \in I$ $\exists e \in I$ $e = e^2$, $\forall k = 1, \ldots, n$ $i_k = i_k e$.

Proof It is obvious that $(3 \Rightarrow 2 \Rightarrow 1)$. It remains to prove $(1 \Rightarrow 3)$. Each i_k has the form

$$i_k = r_1^{(k)} e_1^{(k)} + \cdots + r_{m_k}^{(k)} e_{m_k}^{(k)}$$

with $r_i^{(j)} \in R$, $e_i^{(j)} \in I$, $\left(e_i^{(j)}\right)^2 = e_i^{(j)}$. It suffices to find $e = e^2 \in I$ such that $e_i^{(j)} e = e_i^{(j)}$ for all indices i, j.

This reduces the problem to the following situation. If $e_1, \ldots, e_m$ are idempotent elements in I, there exists an idempotent element $e \in I$ such that for each index i, we have $e_i e = e_i$, that is, $e_i \leq e$. It suffices to choose $e = e_1 \vee \cdots \vee e_n \in I$.   □

Proposition 8.9 *In every ring R*

1. *the ideals $(0) \lhd R$ and $R \lhd R$ are regular;*
2. *the intersection $I \cap J$ of two regular ideals is regular and coincides with the product IJ of these ideals;*
3. *an arbitrary sum $+_{k \in K} I_k$ of regular ideals is still regular.*

Proof The ring itself is regular because $r = r1$ for each $r \in R$, where of course $1 \in R$ is an idempotent element. The ideal (0) is regular because 0 is idempotent.

Given arbitrary ideals I, J we have always $IJ \subseteq I \cap J$. When I and J are regular and $r \in I \cap J$, we can write $r = re$ with $e = e^2 \in J$ (see Proposition 8.8). Thus $r = re$ with $r \in I$ and $e \in J$; this proves that $r \in IJ$ and thus, $I \cap J \subseteq IJ$. So we have already $I \cap J = IJ$.

We must still prove that IJ is regular. Each element of IJ can be written as

$$r = i_1 j_1 + \cdots + i_n j_n, \quad i_k \in I, \ j_k \in J.$$

By Proposition 8.8, there exists an $e = e^2 \in I$ such that $i_k e = i_k$ for all indices k; in the same way there exists an $e' = e'^2 \in J$ such that $j_k e' = j_k$ for all indices k. This yields the idempotent element $e \wedge e' = ee' \in IJ$ with the property that $ree' = r$. Again by Proposition 8.8, the ideal IJ is regular.

Finally let $(I_k)_{k \in K}$ be a family of regular ideals. An element $r \in I = \sum_{k \in K} I_k$ has the form $r = r_1 + \cdots + r_n$ with $r_i \in I_{k_i}$. By Proposition 8.8, for each index i we can choose $e_i = e_i^2 \in I_{k_i} \subseteq I$ such that $r_i e_i = r_i$. We must find $e = e^2 \in I$ such that $e_i e = e_i$, that is, $e_i \leq e$ for each index i. It suffices to choose $e = e_1 \vee \cdots \vee e_n$. □

Regular ideals allow a useful alternative description of the Pierce spectrum of a ring. For this observe first that:

Proposition 8.10 *Let R be a ring. The lattice $\mathsf{Reg}(R)$ of its regular ideals is isomorphic to the lattice of filters of its Boolean algebra of idempotents.*

Proof The expected isomorphism is the following one:

$$\varphi \colon \mathsf{Reg}(R) \longrightarrow \mathsf{Filters}(\mathsf{Idemp}(R)), \quad I \mapsto \{1 - e \in R \mid e = e^2 \in I\}.$$

It is easy to see that each $\varphi(I)$ is a filter:

- $1 \in \varphi(I)$ because $0 \in I$.
- For each pair of idempotent elements $e \in I, e' \in R$, the relation $1 - e \leq e'$ implies $1 - e' = e - ee' \in I$; this forces $e' \in \varphi(I)$.
- For each pair of idempotent elements $e, e' \in I$,

$$(1 - e) \wedge (1 - e') = (1 - e)(1 - e') = 1 - e - e' + ee' = 1 - e'' \in \varphi(I)$$

where $e'' = e + e' - ee' \in I$ is an idempotent element (see Proposition 8.2).

Two regular ideals I, J are equal when they contain the same idempotent elements; thus φ is injective. Moreover, $I \subseteq J$ if and only if $\mathsf{Idemp}(I) \subseteq \mathsf{Idemp}(J)$, thus if and only if $\varphi(I) \subseteq \varphi(J)$. It remains to prove that φ is surjective.

If $F \subseteq \mathsf{Idemp}(R)$ is a filter, consider the regular ideal $I \triangleleft R$ generated by the elements $1 - e$, for each element $e \in F$. This already forces $F \subseteq \varphi(I)$. We must prove that $\varphi(I) \subseteq F$, that is, each element $e = e^2 \in I$ is such that $1 - e \in F$. By definition of I, we can write

$$e = r_1(1 - e_1) + \cdots + r_n(1 - e_n), \quad r_i \in R, \quad e_i \in F.$$

Since F is a filter

$$e' = e_1 \cdots e_n = e_1 \wedge \cdots \wedge e_n \in F.$$

This implies, for each index i,

$$e' \vee e_i = e_i \Rightarrow \complement e' \wedge \complement e_i = \complement e_i \Rightarrow (1 - e')(1 - e_i) = (1 - e_i)$$

and thus as a consequence, $e(1 - e') = e$. Then

$$e(1 - e') = e \Rightarrow e \wedge (1 - e') = e \Rightarrow e \leq \complement e' \Rightarrow e' \leq \complement e \Rightarrow e' \leq (1 - e).$$

Finally $1 - e \in F$ because $e' \in F$ and F is a filter.   □

Corollary 8.11 *Let R be a ring.*

1. *If $M \triangleleft R$ is a maximal regular ideal and $e = e^2 \in R$ is an idempotent element,*

$$e \in M \text{ or } 1 - e \in M.$$

2. *Each regular ideal $I \triangleleft R$ is the intersection of all the maximal regular ideals which contain it.*

Proof By Proposition 8.10 and with the notation of its proof, $\varphi(M)$ is an ultrafilter in the Boolean algebra $\mathsf{Idemp}(A)$. Thus $e \in \varphi(M)$ or $1 - e \in \varphi(M)$ (see Proposition 3.20). The second condition is an immediate consequence of Proposition 3.21.   □

Theorem 8.12 *The Pierce spectrum $\mathsf{Pierce}(R)$ of a ring R is homeomorphic to the following topological space:*

1. *the set of elements is*

$$\mathsf{Sp}(R) = \{M \triangleleft R \,|\, M \text{ maximal regular ideal}\}$$

2. *the open subsets of $\mathsf{Sp}(R)$ are, for each regular ideal $I \triangleleft R$,*

$$\mathcal{U}_I = \{M \in \mathsf{Sp}(R) \,|\, I \not\subseteq M\}.$$

Proof By Proposition 8.10 and with its notation, the isomorphism φ restricts as a bijection between the sets of maximal regular ideals and of that of ultrafilters, thus as a bijection between the two spectra. With the notation of the statement, condition 2 in Corollary 8.11 indicates that two distinct regular ideals I, J determine two distinct open subsets $\mathcal{U}_I, \mathcal{U}_J$, so that $\mathsf{Reg}(R)$ is isomorphic to the lattice of these $\mathcal{U}_I$. By Proposition 3.22 and Definition 8.3, it remains to recall that Proposition 8.10 attests that $\mathsf{Reg}(R)$ is isomorphic to the lattice of filters of the Boolean algebra $\mathsf{Idemp}(R)$, that is, the lattice of open subsets of $\mathsf{Pierce}(R)$.   □

Given an idempotent element $e = e^2 \in R$, the principal ideal $Re \triangleleft R$ is generated by e, thus is regular. Let us simply write $\mathcal{U}_e$ instead of $\mathcal{U}_{Re}$:

$$\mathcal{U}_e = \mathcal{U}_{Re} = \{M \in \mathsf{Sp}(R) \mid Re \not\subseteq M\} = \{M \in \mathsf{Sp}(R) \mid e \notin M\}.$$

Proposition 8.13 *Let R be a ring.*

1. *The open subsets $\mathcal{U}_e \subseteq \mathsf{Sp}(R)$ are exactly the clopens of the spectrum $\mathsf{Sp}(R)$.*
2. *The open subsets $\mathcal{U}_e \subseteq \mathsf{Sp}(R)$ constitute a base of open subsets of $\mathsf{Sp}(R)$.*
3. *Via the isomorphism $\mathsf{Sp}(R) \cong \mathsf{Spec}(\mathsf{Idemp}(R))$ (see Theorem 8.12), the open subset $\mathcal{U}_e \subseteq \mathsf{Sp}(R)$ corresponds to the open subset $O_{1-e} \subseteq \mathsf{Spec}(\mathsf{Idemp}(R))$.*
4. *Given two idempotent elements e and e'*

$$\mathcal{U}_{e \wedge e'} = \mathcal{U}_e \wedge \mathcal{U}_{e'}, \quad \mathcal{U}_{e \vee e'} = \mathcal{U}_e \cup \mathcal{U}_{e'}.$$

Proof Let us first prove condition 3. The idempotents $e' = re$ of Re are such that $e'e = ree = re = e'$, that is $e' \leq e$. Therefore $1 - e \leq 1 - e$ and via the isomorphism in Theorem 8.10, the filter corresponding to Re is the upper segment $\uparrow(1 - e)$. The corresponding open subset in the spectrum of the Boolean algebra $\mathsf{Idemp}(R)$ is then O_{1-e} (see Proposition 3.24).

Condition 4 follows then from Condition 3 and Proposition 3.24. Conditions 1 and 2 are also direct consequences of Theorem 8.12, Corollary 3.26 and Proposition 3.24. □

Proposition 8.14 *Let R be a ring and $e = e^2 \in R$ an idempotent element. Each partition of the clopen $\mathcal{U}_e$ into non-empty clopens has the form*

$$\mathcal{U}_e = \mathcal{U}_{e_1} \cup \cdots \cup \mathcal{U}_{e_n}$$

where

1. *each $e_i \in R$ is a non-zero idempotent element,*
2. *$e_1 + \cdots + e_n = e$,*
3. *$i \neq j \Rightarrow e_i e_j = 0$.*

This result holds in particular for $e = 1$, that is, for $\mathsf{Sp}(R) = \mathcal{U}_1$.

Proof Since $\mathcal{U}_e$ is compact, each partition into non-empty clopens is necessarily finite. By Proposition 8.13, each non-empty clopen has the form $\mathcal{U}_{e'}$ for an idempotent element $e' = e'^2 \neq 0$. A partition of $\mathcal{U}_e$ in non-empty clopens thus has the form

$$\mathcal{U}_e = \mathcal{U}_{e_1} \cup \cdots \cup \mathcal{U}_{e_n}$$

with each e_i idempotent. Still by Proposition 8.13, this corresponds to a partition

$$O_{1-e} = O_{1-e_1} \cup \cdots \cup O_{1-e_n}, \quad O_{1-e_i} \cap O_{1-e_j} = \emptyset \text{ when } i \neq j$$

in the spectrum of the Boolean algebra $\mathsf{Idemp}(R)$. Considering the complements (see Proposition 3.24), we obtain

$$O_e = O_{e_1} \cap \cdots \cap O_{e_n}, \quad O_{e_i} \cup O_{e_j} = \mathsf{Spec}\big(\mathsf{Idemp}(R)\big) \text{ when } i \neq j$$

and thus still by Proposition 3.24

$$e = e_1 \vee \cdots \vee e_n, \quad e_i \wedge e_j = 0 \text{ when } i \neq j. \qquad \square$$

8.3 A Representation Theorem for Rings

In contrast to the case of Boolean algebras (see the Stone duality theorem 3.29), the spectrum of a ring does not characterize that ring: just consider the case of a field, whose spectrum is a singleton (see Proposition 8.31). However, a more involved notion will do the job: the ring will be recaptured as a ring of continuous sections via a topological construction involving the spectrum.

Proposition 8.15 *Let R be a ring.*

1. *If $e_1, \ldots, e_n$ are idempotent elements in R and $\langle e_1, \ldots, e_n \rangle$ indicates the regular ideal generated by these, then*

$$\langle e_1, \ldots, e_n \rangle = \langle e_1 \vee \cdots \vee e_n \rangle$$

where $e_1 \vee \cdots \vee e_n$ is the supremum of the elements e_i in the Boolean algebra Idemp(R).
2. *If $e \in R$ is an idempotent element,*

$$R/\langle e \rangle \cong R(1 - e).$$

Proof If e, e' are idempotent elements in an ideal I

$$e \vee e' = e + e' - ee' \in I.$$

Iterating this argument, we get

$$e_1 \vee \cdots \vee e_n \in \langle e_1, \ldots, e_n \rangle$$

thus

$$\langle e_1 \vee \cdots \vee e_n \rangle \subseteq \langle e_1, \ldots, e_n \rangle.$$

Conversely $e_i \leq e_1 \vee \cdots \vee e_n$, where

$$e_i = e_i \wedge (e_1 \vee \cdots \vee e_n) = e_i(e_1 \vee \cdots \vee e_n) \in \langle e_1 \vee \cdots \vee e_n \rangle$$

and so

$$\langle e_1, \ldots, e_n \rangle \subseteq \langle e_1 \vee \cdots e_n \rangle.$$

To prove condition 2, observe the existence of the direct sum

$$R \cong Re \oplus R(1 - e).$$

Indeed, $R = Re + R(1 - e)$ since $1 = e + (1 - e)$. Moreover by Proposition 8.9, $Re \cap R(1 - e) = ReR(1 - e) = (0)$ because $e(1 - e) = e - e^2 = 0$. Thus $R/Re \cong R(1 - e)$.   □

Proposition 8.16 *Let $M \vartriangleleft R$ be a maximal regular ideal of a non-zero ring R. The quotient R/M admits 0 and 1 as its only idempotent elements.*

Proof The ideal M is the filtered union of all the regular ideals $\langle e_1, \ldots, e_n \rangle$ generated by finitely many idempotent elements of M. The quotient R/M is thus the filtered colimit of the quotients $R/\langle e_1, \ldots, e_n \rangle$. Choose $r \in R$ such that

$$[r] = [r]^2 \in R/M = \operatorname*{colim}_{e_1, \ldots, e_n} R/\langle e_1, \ldots, e_n \rangle$$

in the filtered colimit of quotients. That filtered colimit is constructed as in the category of sets, thus there exists a term of the colimit such that $[r] = [r]^2 \in R/\langle e_1, \ldots, e_n \rangle$. Applying Proposition 8.15 to the idempotent element, $e = e_1 \vee \cdots \vee e_n$

$$[r] = [r]^2 \in R/\langle e_1, \ldots, e_m \rangle = R/Re \cong R(1 - e), \quad e = e_1 \vee \cdots \vee e_n.$$

Let $e' \in R(1 - e)$ be the idempotent element corresponding to r under this isomorphism. We have thus $[r] = [e'] \in R/Re$ and this implies further $[r] = [e'] \in R/M$. If $e' \in M$, then $[e'] = 0 \in R/M$; if $e' \notin M$, then $1 - e' \in M$ (see Corollary 8.11) and $[e'] = 1 \in R/M$. □

Definition 8.17 The structural space of a ring R is the disjoint union of all the quotients R/M, for all the maximal regular ideals $M \lhd R$, provided with the final topology[1] for all the mappings

$$s_r^I : \mathcal{U}_I \longrightarrow \coprod_M R/M, \quad N \mapsto [r] \in R/N$$

where I runs through the regular ideals $I \lhd R$ and r runs through the elements of R.

Let us recall a classical notion of general topology:

Definition 8.18 A mapping $f : X \longrightarrow Y$ between topological spaces is *étale* when, for each point $x \in X$, there exist open neighborhoods U_x of x and V_x of $f(x)$ such that f restricts to a homeomorphism $f : U_x \longrightarrow V_x$.

The classical example of an étale mapping is the projection of a circular helix onto a base circle.

Proposition 8.19 *An étale mapping between topological spaces is continuous and open.*

Proof By definition, an étale mapping $f : X \longrightarrow Y$ is continuous on a neighborhood of each point of X, thus is continuous.

Next, if $U \subseteq X$ is open, each point $x \in U$ admits an open neighborhood U_x with the property indicated in Definition 8.18

$$f : U_x \overset{\cong}{\longrightarrow} V_x,$$

and there is no restriction in supposing $U_x \subseteq U$. The open subset U is then the union $U = \bigcup_{x \in U} U_x$ and thus

$$f(U) = \bigcup_{x \in U} f(U_x) = \bigcup_{x \in U} V_x$$

is open, as a union of open subsets. □

---

[1] Given a family of mappings $(f_i : X_i \longrightarrow Y)_{i \in I}$ with each X_i a topological space and Y a set, the corresponding final topology on Y has for open subsets those $U \subseteq Y$ such that $f_i^{-1}(U) \subseteq X_i$ is open, for each index i.

Theorem 8.20 *Let R be a ring. The mapping*

$$p: \coprod_M R/M \longrightarrow \mathsf{Sp}(R), \quad [r] \in R/N \mapsto N$$

is étale.

Proof Let us consider an element $r \in R$ and prove first that

$$U_r = \{M \in \mathsf{Sp}(R) | r \in M\}$$

is open in $\mathsf{Sp}(R)$.

For this consider

$$J = \{e = e^2 \in R | \forall N \in \mathsf{Sp}(R) \ r \notin N \Rightarrow e \in N\}.$$

Let us write $I \triangleleft R$ for the regular ideal generated by all the elements of J. Observe that $J = \mathsf{Idemp}(I)$. The inclusion $J \subseteq \mathsf{Idemp}(I)$ is obvious. The other inclusion is easy: if $e = e^2 \in I$,

$$e = r_1 e_1 + \cdots + r_n e_n, \quad e_1, \ldots, e_n \in J.$$

If $N \in \mathsf{Sp}(R)$ and $r \notin N$, we have $e_i \in N$ for each index i, thus $e \in N$.

Now for each $N \in \mathsf{Sp}(R)$, we have

$$\begin{aligned}
N \in \mathcal{U}_I &\Leftrightarrow I \nsubseteq N \\
&\Leftrightarrow \exists e \ e = e^2, \ e \in I, \ e \notin N \\
&\Leftrightarrow \exists e \ e \in J, \ e \notin N \\
&\Rightarrow r \in N \\
&\Leftrightarrow N \in U_r.
\end{aligned}$$

Observe further that the only implication in the formulæ above is an equivalence as well; this will prove $\mathcal{U}_I = U_r$. Indeed if $r \in N$, by Proposition 8.8 we get $r = re'$ with $e' = e'^2 \in N$. If $M \in \mathsf{Sp}(R)$ and $r \notin M$, we have $e' \notin M$ because $r = re'$. Then $1 - e' \in M$ by maximalilty of M (see Corollary 8.11). This proves that $e = 1 - e' \in J$ with $e = 1 - e' \notin N$, because $e' \in N$. This concludes the proof of the equality $U_r = \mathcal{U}_I$; thus U_r is open.

Now if s_r^I and $s_t^{I'}$ are two sections of p as in Definition 8.17:

$$\begin{aligned}
(s_t^{I'})^{-1}(s_r^I(\mathcal{U}_I)) &= \{M \in \mathcal{U}_{I'} | s_t^{I'}(M) \in s_r^I(\mathcal{U}_I)\} \\
&= \{M \in \mathcal{U}_I \cap \mathcal{U}_{I'} | [t] = [r] \in R/M\} \\
&= \{M \in \mathcal{U}_{I \cap I'} | [t - r] = 0 \in R/M\} \\
&= \{M \in \mathcal{U}_{I \cap I'} | t - r \in M\} \\
&= \mathcal{U}_{I \cap I'} \cap U_{t-r}
\end{aligned}$$

and this last term is an open subset. By definition of a final topology, this implies that $s_r^I(\mathcal{U}_I)$ is open in $\coprod_M R/M$.

Again by definition of a final topology, the continuity of p is equivalent to the continuity of all the composites $p \circ s_r^I$. These composites are the inclusions of the open subsets $\mathcal{U}_I$ in $\mathsf{Sp}(R)$, thus are continuous.

Finally, for each element $[r] \in R/M$, we have inverse homeomorphisms

$$\mathsf{Sp}(R) \underset{s_r^R}{\overset{p}{\rightleftarrows}} s_r^R(\mathsf{Sp}(R))$$

with $s_r^R(\mathsf{Sp}(R))$ an open neighborhood of $[r]$. $\qquad\square$

Corollary 8.21 *With the notation of Definition 8.17, each "fibre" R/M is a discrete subspace of the structural space of the ring R.*

Proof For each $r \in R$, the proof of Theorem 8.20 has shown that $s_r^R(\mathsf{Sp}(R))$ is an open subset of the structural space, homeomorphic to $\mathsf{Sp}(R)$. Its trace on the fibre R/M is the singleton $[r]$. $\qquad\square$

Lemma 8.22 *Let R be a ring and*

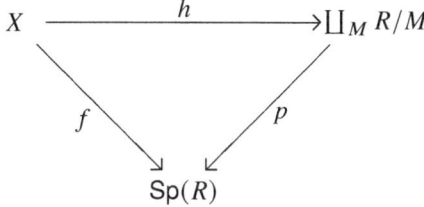

a commutative triangle of continuous mappings, with p the étale mapping of Theorem 8.20. When X is a profinite space, there exists a finite partition

$$X = U_1 \cup \cdots \cup U_n,$$

with each $U_i \subseteq X$ clopen, and elements

$$r_1, \ldots, r_n \in R$$

such that

$$h = \sum_{i=1}^{n} r_i \chi_{U_i}$$

where, for each clopen $U_i \subseteq X$, χ_{U_i} is the continuous mapping

$$\chi_{U_i} : X \longrightarrow \coprod_M R/M, \quad \begin{cases} x \mapsto [1] \in R/f(x) \text{ if } x \in U_i, \\ x \mapsto [0] \in R/f(x) \text{ if } x \notin U_i. \end{cases}$$

Proof With the notation of Definition 8.17, the subsets

$$h^{-1}\left(s_r^{\mathsf{Sp}(R)}\left(\mathsf{Sp}(R)\right)\right),$$

for each $r \in R$, constitute an open covering of X (see the proof of Theorem 8.20). Since X has a base of clopens (see Corollary 3.14), there exists a refinement of that covering, constituted of clopens. But since X is compact, we can next choose a finite subcovering constituted of clopens. Starting with the first clopen, adding the second one minus its intersection with the first one, then the third one minus its intersections with the first two ones, and so on, we obtain this time a finite partition constituted of clopens. We have thus obtained a partition into clopens $U_1, \ldots, U_n \subseteq X$ together with elements $r_1, \ldots, r_n \in R$ such that for each index i and each element $x \in U_i$, $h(x) = [r_i] \in R/f(x)$. This proves that $h = \sum_{i=1}^n r_i \cdot \chi_{U_i}$.

Observe that on each clopen U_i, χ_{U_i} is the restriction of the continuous mapping $s_1^R \circ f$ of Definition 8.17, while for $j \neq i$, χ_{U_j} is the restriction of the continuous mapping $s_0^R \circ f$. This proves the continuity of each χ_{U_i}.   □

Theorem 8.23 *Let R be a ring.*

1. *R is isomorphic to the ring of continuous sections of the étale mapping*

$$p: \coprod_M R/M \longrightarrow \mathsf{Sp}(R), \quad [r] \in R/N \mapsto N,$$

   of its structural space.
2. *For each idempotent element $e = e^2 \in R$, the ring of continuous sections of p on the open subset $\mathcal{U}_e$ of $\mathsf{Sp}(R)$ is isomorphic to the principal regular ideal Re.*

Proof Let us fix an idempotent element $e \in R$; the first part of the statement is the special case $e = 1$. For each element $r \in R$ we have, by definition of the topology of the structural space, a continuous section

$$s_r^e = s_r^{Re}: \mathcal{U}_e \longrightarrow \coprod_M R/M, \quad N \mapsto [r] \in R/N.$$

Let us write $\mathsf{Sec}_e(p)$ for the set of continuous sections of p on $\mathcal{U}_e$ and let us consider the function

$$\varphi_e: R \longrightarrow \mathsf{Sec}_e(p), \quad r \mapsto s_r^e.$$

Let us first observe that the continuous functions $\varphi_e(r)$ constitute a ring for the pointwise operations in each "fibre" of the structural space. Indeed

$$\left(\varphi_e(r) + \varphi_e(r')\right)(N) = [r] + [r'] = [r + r'] = \varphi_e(r + r')(N) \in R/N$$

and analogously for the multiplication. Trivially φ_e preserves the ring operations:

$$\left(\varphi_e(r) + \varphi_e(s)\right)(M) = \varphi_e(r)(M) + \varphi_e(s)(M) = [r] + [s] = [r + s] = \varphi_e(r + s)(M)$$

and analogously for the multiplication. Thus φ_e is a ring homomorphism.

Observe now that by Proposition 8.11,

$$M \in \mathcal{U}_e \Leftrightarrow e \notin M \Leftrightarrow 1 - e \in M;$$

thus $s^e_{1-e} = 0$ because $0 = [1-e] \in R/M$. This implies the existence of a factorization of φ_e

$$R \twoheadrightarrow R/R(1-e) \xrightarrow[\psi_e]{} \varphi_e(R), \qquad \psi_e([r]) = \varphi_e(r).$$

This factorization ψ_e is injective because when $s^e_r = 0$,

$$r \in \bigcap_{M \in \mathcal{U}_e} M = \bigcap_{e \notin M} M = \bigcap_{1-e \in M} M = \bigcap_{R(1-e) \subseteq M} M = R(1-e)$$

by Corollary 8.11. The factorization ψ_e is also surjective, by definition of its codomain. By Proposition 8.15, we obtain an isomorphism

$$Re \cong R/R(1-e) \xrightarrow{\cong} \varphi_e(R).$$

It remains to prove the surjectivity of φ_e. For this, choose a continuous section $\sigma \in \mathrm{Sec}_e(p)$. By Lemma 8.22 and condition 8.13.1, there exist idempotent elements $e_1, \ldots, e_n \in R$ and arbitrary elements $r_1, \ldots, r_n \in R$ such that

$$\mathcal{U}_e = \mathcal{U}_{e_1} \cup \cdots \cup \mathcal{U}_{e_n}, \quad \sigma = \sum_{i=1}^{n} r_i \chi_{\mathcal{U}_{e_i}}.$$

Let us put $r = r_1 e_1 + \cdots + r_n e_n$ and prove that $\sigma = s^e_r$. If $M \in \mathcal{U}_{e_i}$, we get $e_i \notin M$, thus $1 - e_i \in M$ (see Corollary 8.11). Since we are working with a partition, we also have $M \notin \mathcal{U}_{e_j}$ for $j \neq i$, that is, $e_j \in M$. As a consequence, in R/M we get:

$$
\begin{aligned}
s^e_r(M) &= [r_1 e_1 + \cdots + r_n e_n] \\
&= [r_1][e_1] + \cdots + [r_n][e_n] \\
&= [r_i][e_i] \quad \text{because } e_j \in M \text{ for } j \neq i \\
&= [r_i] \quad \text{because } 1 - e_i \in M, \text{ thus } [e_i] = [1] \in R/M \\
&= \sigma(M).
\end{aligned}
$$

This concludes the proof. $\qquad \Box$

Theorem 8.23 allows us to study the ring R via a "bunch" of rings R/M whose spectra are just singletons, by Proposition 8.16.

8.4 The Fibred Spectrum of an Algebra

Of course an R-algebra is in particular a ring, but its spectrum – as a ring – is also related to the spectrum of the ring R:

Definition 8.24 Let R be ring. For each R-algebra A, consider the morphism of R-algebras

$$\rho_A \colon R \longrightarrow A, \quad r \mapsto r \cdot 1.$$

The fibred spectrum of the R-algebra A is the object

$$\Big(\mathsf{Sp}(\rho_A) \colon \mathsf{Sp}(A) \longrightarrow \mathsf{Sp}(R)\Big)$$

of $\mathsf{Prof}/\mathsf{Sp}(R)$.

Proposition 8.25 *Let R be a ring. The construction of the fibred spectrum of an R-algebra extends as a contravariant functor*

$$\mathsf{Sp}_R \colon \mathsf{Alg}_R \longrightarrow \mathsf{Prof}/\mathsf{Sp}(R).$$

Proof When $f \colon A \longrightarrow B$ is a morphism of R-algebras, we have $f \circ \rho_A = \rho_B$. Thus

$$\mathsf{Sp}(f) \colon \big(\mathsf{Sp}(\rho_B) \colon \mathsf{Sp}(B) \to \mathsf{Sp}(R)\big) \longrightarrow \big(\mathsf{Sp}(\rho_A) \colon \mathsf{Sp}(A) \to \mathsf{Sp}(R)\big)$$

is a morphism in $\mathsf{Prof}/\mathsf{Sp}(R)$. $\square$

The following property of the spectrum functor will be essential to get the Galois theorem for rings.

Theorem 8.26 *For each ring R, the functor*

$$\mathsf{Sp}_R \colon \mathsf{Alg}_R^{\mathrm{op}} \longrightarrow \mathsf{Prof}/\mathsf{Sp}(R), \quad A \mapsto \big(\mathsf{Sp}(A) \longrightarrow \mathsf{Sp}(R)\big)$$

has a full and faithful right adjoint functor

$$C_R \colon \mathsf{Prof}/\mathsf{Sp}(R) \longrightarrow \mathsf{Alg}_R^{\mathrm{op}}, \quad (X, f) \mapsto \mathsf{Cont}\Big((X, f), \big(\textstyle\coprod_M R/M, p\big)\Big),$$

where

- $p \colon \coprod_M R/M \longrightarrow \mathsf{Sp}(R)$ *is the canonical morphism of the structural space of the ring R (see Definition 8.17);*
- $\mathsf{Cont}\Big((X, f), \big(\coprod_M R/M, p\big)\Big) = \left\{ h \colon X \longrightarrow \coprod_M R/M \;\middle|\; \begin{array}{l} h \text{ continuous} \\ p \circ h = f \end{array} \right\};$
- $\mathsf{Cont}\Big((X, f), \big(\coprod_M R/M, p\big)\Big)$ *has the structure of an R-algebra induced pointwise by the structure of each R-algebra R/M.*

In particular, the co-unit of the adjunction is an isomorphism.

Proof The case $R = \{0\}$ is obvious because $\mathsf{Alg}_R^{\mathrm{op}} \cong \{(0)\}$ and $\mathsf{Prof}/\mathsf{Sp}(R) \cong \{\emptyset \Longrightarrow \emptyset\}$. Let us thus suppose that $0 \neq 1$ in R.

First, observe that the set $\mathsf{Cont}\big((X, f), (\coprod_M R/M, p)\big)$ can be provided with the structure of an R-algebra, defined pointwise via the R-algebra structure of each fibre R/M. Clearly the two mappings

$$x \mapsto [0] \in R/f(x), \quad x \mapsto [1] \in R/f(x)$$

are continuous, since they are just the composites $s_0^R \circ f$ and $s_1^R \circ f$ (see Definition 8.17). We must next prove that

$$g, g' \text{ continuous} \Rightarrow g + g', \ g - g', \ g \times g' \text{ continuous.}$$

Let us do the job for $g + g'$; the other cases are analogous. By Lemma 8.22 and with its notation, there exists a partition into clopens $X = U_1 \cup \ldots \cup U_n$ and elements $r_i \in R$ such that $g = \sum_{i=1}^n r_i \chi_{U_i}$; analogously there exists a partition into clopens $X = U_1' \cup \ldots \cup U_m'$ and elements $r_j' \in R$ such that $g' = \sum_{j=1}^m r_j' \chi_{U_j'}$. The $U_i \cap U_j'$ constitute a partition into clopens such that $g + g = \sum_{i,j}(r_i + r_j')\chi_{U_i \cap U_j'}$ coincides on $U_i \cap U_j$ with the restriction of the continuous mapping $s_{r_i + r'j}^R \circ f$ (see Definition 8.17), thus is continuous.

Let us now construct the unit of the adjunction. Let A be an R-algebra. For each continuous mapping

$$h: \big(\mathsf{Sp}(A), \mathsf{Sp}(\rho_A)\big) \longrightarrow \big(\textstyle\coprod_M R/M, p\big)$$

there exist idempotent elements $e_1, \ldots, e_n \in A$ and elements $r_1, \ldots, r_n \in R$ such that $h = \sum_{i=1}^n r_i \chi_{\mathcal{U}_{e_i}}$ (see Lemma 8.22 and Proposition 8.13). Let us define

$$\alpha_A: C_R \mathsf{Sp}_R(A) \longrightarrow A, \quad h \mapsto \sum_{i=1}^n r_i e_i.$$

We must prove that this definition is independent of the decomposition $h = \sum_{i=1}^n r_i \chi_{\mathcal{U}_{e_i}}$. Then the naturality of α will be obvious.

Let us choose another decomposition $h = \sum_{j=1}^m r_j' \chi_{\mathcal{U}_{e_j'}}$ as above. Considering

$$\mathcal{U}_{e_i} \cap \mathcal{U}_{e_j'} = \mathcal{U}_{e_i e_j'}$$

(see Proposition 8.13) we get a new partition of $\mathsf{Sp}(A)$ into clopens, with the property

$$x \in \mathcal{U}_{e_i e_j'} \Rightarrow [r_i] = h(x) = [r_j'] \in R/f(x).$$

But, when fixing an index i, the clopens $\mathcal{U}_{e_i e_j'}$ for $j = 1, \ldots, m$ constitute a finite partition of $\mathcal{U}_{e_i}$, still into clopens. By Lemma 8.14, this implies $e_i = \sum_{j=1}^m e_i e_j'$. As a consequence, we get

$$r_i e_i = r_i \sum_{j=1}^{m} e_i e'_j$$

and in an analogous way

$$r'_j e'_j = r'_j \sum_{i=1}^{n} e_i e'_j.$$

It suffices now to prove that $r_i e_i e'_j = r'_j e_i e'_j$, that is

$$(r_i - r'_j) e_i e'_j = 0$$

for each pair of indices i, j: this will show that the definition of α_A is independent of the decomposition of h. But for $M \in \mathcal{U}_{e_i e'_j}$, the equality $[r_i] = [r'_j] \in R/M$ implies $r_i - r'_j \in M$. It thus suffices to prove that for each idempotent element $e \in A$ (here, $e = e_i e'_j$) and for each element $r \in R$ (here, $r = r_i - r'_j$)

$$(\forall M \in \mathcal{U}_e \ \ r \in M) \Rightarrow (re = 0).$$

In the proof of Theorem 8.23, we have seen that

$$(\forall M \in \mathcal{U}_e \ \ r \in M) \Rightarrow (r \in R(1 - e)),$$

thus via the isomorphism $Re \cong R/(1 - e)R$ (see Proposition 8.15), $re = 0$.

Next, let us construct the co-unit of the adjunction. Each mapping χ_U in Lemma 8.22 is an idempotent element in $C_R(X, f)$. This yields a homomorphism

$$\beta' : \mathsf{Clopen}(X) \longrightarrow \mathsf{Idemp}(C_R(X, f)), \quad U \mapsto \beta'(U) = \chi_U,$$

and this homomorphism corresponds by the Stone duality to a continuous mapping $\beta : \mathsf{Sp}(C_R(X, f)) \longrightarrow X$. Let us simply write

$$\rho : \mathsf{Sp}(C_R(X, f)) \longrightarrow \mathsf{Sp}(R)$$

for the morphism $\mathsf{Sp}(\rho_{C_R(X,f)})$ of Definition 8.24. To get a morphism

$$\beta : \Big(\mathsf{Sp}(C_R(X, f)), \rho\Big) \longrightarrow (X, f),$$

we must still prove the equality $f \circ \beta = \rho$. This means, for each idempotent element $e \in R$, the equality $\beta'(f^{-1}(O_e)) = e \cdot 1$. We have

$$\chi_{f^{-1}(O_e)}(x) = \begin{cases} [1] & \text{if } x \in f^{-1}(O_e), \text{ i.e. } e \notin f(x), \\ [0] & \text{if } x \notin f^{-1}(O_e), \text{ i.e. } e \in f(x). \end{cases}$$

But $e \notin f(x)$ implies $1 - e \in f(x)$, thus $[1] = [e] \in R/f(x)$; moreover $e \in f(x)$ implies $[0] = [e] \in R/f(x)$. This proves that $\chi_{f^{-1}(O_e)}(x) = [e] = e \cdot [1]$ for each element $x \in X$. Now

$$\beta'\big(f^{-1}(O_e)\big) = \chi_{f^{-1}(O_e)} = e \cdot 1 = e.$$

It is obvious that β is a natural transformation.

Let us prove now that this co-unit β is a homeomorphism; this will imply that the functor C_R is full and faithful (see Proposition 5.5). An idempotent element of the ring $C_R(X, f)$ is a continuous mapping $g\colon X \longrightarrow \coprod_M R/M$ such that $p \circ g = f$ with each $g(x) \in R/f(x)$ idempotent; that is, each $g(x)$ is equal to 0 or 1, by Proposition 8.16. The sets

$$s_0^{\mathsf{Sp}(R)}\big(\mathsf{Sp}(R)\big) = \big\{[0] \in R/M \,\big|\, M \in \mathsf{Sp}(R)\big\},$$
$$s_1^{\mathsf{Sp}(R)}\big(\mathsf{Sp}(R)\big) = \big\{[1] \in R/M \,\big|\, M \in \mathsf{Sp}(R)\big\}$$

are open (see the proof of Theorem 8.20) and disjoint (because $1 \notin M$, which is a proper ideal). But the idempotent elements of $C_R(X, f)$ are

$$\mathsf{Cont}\Big((X, f), \big(\{0, 1\} \times \mathsf{Sp}(R), p_2\big)\Big),$$

where $\{0, 1\}$ has the discrete topology and p_2 is the projection onto the factor $\mathsf{Sp}(R)$. An element of this set is the unique factorization through the product of a pair (u, f), where $u\colon X \longrightarrow \{0, 1\}$ is a continuous mapping. This implies that the set above is isomorphic to

$$C(X, \{0, 1\}) \cong \mathsf{Clopen}(X),$$

the Boolean algebra of clopens of X. By the Stone duality (see Theorem 3.29) we get

$$\mathsf{Sp}\Big(\mathsf{Cont}\big((X, f), (\coprod_M R/M, p)\big)\Big) \cong X.$$

Finally let us prove the triangular identities of the adjunction (see Proposition 5.4). Via the Stone duality again (see Theorem 3.29), for each R-algebra A, the equality

$$\beta_{\mathsf{Sp}_R(A)} \circ \alpha_A = \mathsf{id}_{\mathsf{Sp}_R(A)}$$

becomes, for each idempotent element $e = e^2 \in A$,

$$\alpha_A(\chi_{u_e}) = e,$$

which is trivial.

Next, if $(X, f) \in \mathsf{Prof}/\mathsf{Sp}(R)$, the second equality for the adjunction means

$$\alpha_{C_R(X,f)} \circ C_R\big(\beta_{(X,f)}\big) = \mathsf{id}_{C_R(X,f)}.$$

But each element $h \in C_R(X, f)$ has the form $h = \sum_{i=1}^{n} r_i \chi_{u_i}$, by Lemma 8.22. Since we are considering morphisms of R-algebras, it suffices to prove that for each clopen $U \subseteq X$,

$$\Big(\alpha_{C_R(X,f)} \circ C_R\big(\beta_{(X,f)}\big)\Big)(\chi_U) = U.$$

That is,

$$\alpha_{C_R(X,f)}\left(\chi_U \circ \beta_{(X,f)}\right) = \chi_U.$$

Since χ_U takes only the values 0 and 1, the composite $\chi_U \circ \beta_{(X,f)}$ also takes only the values 0 and 1. Since $\beta_{(X,f)}$ is a homeomorphism and χ_U takes the value 1 exactly on U, the definition of $\beta_{(X,f)}$ implies that $\chi_U \circ \beta_{(X,f)}$ takes the value 1 exactly on the open subset $\mathcal{U}_{\chi_U} \subseteq \mathsf{Sp}\big(C_R(X,f)\big)$. By definition of $\alpha_{C_R(X,f)}$, this forces the conclusion.   □

Corollary 8.27 *The "spectrum functor"*

$$\mathsf{Sp}: \mathsf{Rings}^{\mathsf{op}} \longrightarrow \mathsf{Prof}$$

admits a full and faithful right adjoint functor

$$C(-,\mathbb{Z}): \mathsf{Prof} \longrightarrow \mathsf{Rings}^{\mathsf{op}}$$

where the ring $\mathbb{Z}$ of integers is provided with the discrete topology and $C(X,\mathbb{Z})$ denotes the ring of continuous functions, with the pointwise operations.

Proof A $\mathbb{Z}$-algebra is just a ring. The ring $\mathbb{Z}$ has 0 and 1 as its only idempotents, thus admits (0) as its only maximal regular ideal. Its spectrum is thus a singleton. The structural space of $\mathbb{Z}$ is then $\mathbb{Z}/(0) \cong \mathbb{Z}$, provided with the discrete topology, as attested by Corollary 8.21. We conclude by Theorem 8.26.   □

Here is now a more formal approach to the fibred spectrum of an R-algebra.

Lemma 8.28 *Let $L \dashv R$ be a pair of adjoint functors*

$$L: \mathcal{A} \longrightarrow \mathcal{B}, \quad R: \mathcal{B} \longrightarrow \mathcal{A}.$$

When the category $\mathcal{A}$ has pullbacks, for each object $A \in \mathcal{A}$, the functor

$$L_A: \mathcal{A}/A \longrightarrow \mathcal{B}/L(A), \quad (f: M \to A) \mapsto \big(L(f): L(M) \to L(A)\big)$$

admits the right adjoint functor

$$R_A: \mathcal{B}/L(A) \longrightarrow \mathcal{A}/A, \quad (g: N \to L(A)) \mapsto (N', g')$$

where (N', g') is defined by the pullback

$$
\begin{array}{ccc}
N' & \xrightarrow{\ g''\ } & R(N) \\
{\scriptstyle g'}\big\downarrow & & \big\downarrow{\scriptstyle R(g)} \\
A & \xrightarrow[\ \varepsilon_A\ ]{} & RL(A)
\end{array}
$$

and ε_A is the unit of the adjunction $L \dashv R$.

Proof It is obvious how to extend the construction of the statement to a functor R_A. To prove the adjunction, observe that for each morphism $f \colon D \longrightarrow A$ in $\mathcal{A}$, we have

$$
\begin{aligned}
\mathcal{A}/A\big((D, f), R_A(N, g)\big) &\cong \{h \colon D \to N' \,|\, g' \circ h = h\} \\
&\cong \{s \colon D \to R(N) \,|\, R(g) \circ s = \varepsilon_A \circ f\} \\
&\cong \{t \colon L(K) \to N \,|\, g \circ t = L(f)\} \\
&\cong \mathcal{B}/L(A)\big(L_A(D, f), (N, g)\big)
\end{aligned}
$$

by definition of a pullback and an adjunction.   □

Corollary 8.29 *Let R be a ring. The right adjoint of the functor*

$$
\mathsf{Sp}_R \colon \mathsf{Alg}_R \longrightarrow \mathsf{Prof}/\mathsf{Sp}(R).
$$

is isomorphic to the functor

$$
\mathsf{Prof}/\mathsf{Sp}(R) \longrightarrow \mathsf{Alg}_R, \quad (X, f) \mapsto C_R(X, f) \cong R \otimes_{C\left(\mathsf{Sp}(R), \mathbb{Z}\right)} C(X, \mathbb{Z}).
$$

Proof It suffices to apply Lemma 8.28 to the adjunction of Corollary 8.27, choosing as object A the ring R itself. Indeed, in the category of rings we have the pushout

$$
\begin{array}{ccc}
R \otimes_{C\left(\mathsf{Sp}(R), \mathbb{Z}\right)} C(X, \mathbb{Z}) & \longleftarrow & C(X, \mathbb{Z}) \\
\uparrow & & \uparrow{\scriptstyle C(f, \mathsf{id}_{\mathbb{Z}})} \\
R & \longleftarrow & C\big(\mathsf{Sp}(R), \mathbb{Z}\big)
\end{array}
$$

where the bottom morphism is the unit of the adjunction in Corollary 8.27 (see Proposition 7.24; the rings are the $\mathbb{Z}$-algebras).   □

8.5 The Case of a Field

In the case of a field:

Lemma 8.30 *The only idempotent elements of a field K are 0 and 1.*

Proof If $e = e^2$ with $e \neq 0$ in a field K, we can divide $ee = e1$ by e to get $e = 1$. □

Proposition 8.31 *The Pierce spectrum of a field is a singleton.*

Proof The Boolean algebra of idempotents is $\{0, 1\}$, thus admits $\{0\}$ as its only ultrafilter.   □

Proposition 8.32 *The structural space of a field K is K itself provided with the discrete topology.*

Proof By Proposition 8.31 and Corollary 8.21.   □

Proposition 8.33 *Let K be a field. The functor "spectrum of a K-algebra"*

$$\mathsf{Sp}\colon \mathsf{Alg}_K^{\mathrm{op}} \longrightarrow \mathsf{Prof}$$

admits the full and faithful right adjoint functor

$$C(-, K)\colon \mathsf{Prof} \longrightarrow \mathsf{Alg}_K^{\mathrm{op}}$$

where the field K is provided with the discrete topology and $C(X, K)$ indicates the K-algebra of continuous mappings, with the pointwise operations.

Proof By Proposition 8.31, $\mathsf{Prof}/\mathsf{Sp}(K) \cong \mathsf{Prof}$. The conclusion follows from Proposition 8.32 and Theorem 8.26.   □

Corollary 8.34 *Let $K \subseteq L$ be a field extension. For each profinite space X, there exists an isomorphism of L-algebras*

$$L \otimes_K C(X, K) \cong C(X, L).$$

Proof Using the adjunction in Proposition 2.10 and Proposition 8.33, we get, for each profinite space X and each L-algebra B,

$$
\begin{aligned}
\mathsf{Hom}_L\big(L \otimes_K C(X, K), B\big) &\cong \mathsf{Hom}_K\big(C(X, K), B\big) \\
&\cong \mathsf{Cont}\big(\mathsf{Sp}(B), X\big) \\
&\cong \mathsf{Hom}_L\big(C(X, L), B\big),
\end{aligned}
$$

because the spectrum of B as an L-algebra or as a K-algebra is the same, namely, the spectrum of the ring B.   □

The following result gives evidence that, even if the spectrum of a field is trivial, the theory of the spectrum is well present in the Galois theory of fields.

Proposition 8.35 *Let $K \subseteq L$ be a Galois extension of fields. The Galois equivalence of Theorem 4.24*

$$\mathsf{Hom}_K(-, L)\colon \mathsf{Split}[L : K] \longrightarrow \mathsf{Gal}[L : K]\text{-}\mathsf{Prof}$$

is isomorphic to the functor

$$\mathsf{Sp}(L \otimes_K -)\colon \mathsf{Split}[L : K] \longrightarrow \mathsf{Gal}[L : K]\text{-}\mathsf{Prof}.$$

In particular, when A is finite-dimensional, $\mathsf{Sp}(L \otimes_K A)$ is a finite discrete space.

Proof When the K-algebra A is finite-dimensional, we know by Proposition 4.21 that $\mathsf{Hom}_K(A, L) \cong \mathsf{Hom}_K(A, M)$ for a finite-dimensional Galois extension $K \subseteq M \subseteq L$. Moreover, by Theorem 2.27, $\#\mathsf{Hom}_K(A, M) = n = \dim A$. A finite profinite space is discrete, thus $\mathsf{Hom}_K(A, L)$ is the discrete space with n points.

Moreover when A is split by M, A is trivially split by each extension containing M. Thus by Proposition 4.3, L is the filtered union of all the finite-dimensional Galois subextensions $K \subseteq M \subseteq L$ such that A is split by M. This implies that L^n is the filtered union of the powers M^n, because if $(l_i)_i \in L^n$ is such that each l_i is in some M_i, all components l_i are in the same M. Finally,

$$A \otimes_K L \cong A \otimes_K \left(\bigcup_M M \right) \cong \bigcup_M A \otimes_K M \cong \bigcup_M M^n \cong L^n$$

because $- \otimes_K A$ preserves colimits (see Example 5.2 and Proposition 5.6).

But $\mathsf{Sp}(L^n)$ is also the discrete space with n points, because L admits only the idempotents 0 and 1. Indeed an idempotent element of L^n is a family $(e_1, \ldots, e_n)$ where $e_i = 0$ or $e_i = 1$ for each index i. Identifying such an element with the subset of those indices for which $e_i = 1$, we get that $\mathsf{Idemp}(L^n)$ is isomorphic to the Boolean algebra $\wp(n)$ of subsets of a set with n elements. By the Stone duality (see Theorem 3.29), $\mathsf{Sp}(L^n)$ is a discrete space with n points.

Next when A has arbitrary dimension, Proposition 4.22 implies

$$\mathsf{Hom}_K(A, L) \cong \lim_B \mathsf{Hom}_K(B, L),$$

where B runs through the finite-dimensional subalgebras of A. Applying Proposition 8.5, we obtain

$$\begin{aligned}
\mathsf{Sp}(A \otimes_K L) &\cong \mathsf{Sp}\big((\mathrm{colim}_B B) \otimes_K L\big) \\
&\cong \mathsf{Sp}\big(\mathrm{colim}_B (B \otimes_K L)\big) \\
&\cong \lim_B \mathsf{Sp}(B \otimes_K L) \\
&\cong \lim_B \mathsf{Hom}_K(B, L) \\
&\cong \mathsf{Hom}_K(\mathrm{colim}_B B, L) \\
&\cong \mathsf{Hom}_K(A, L).
\end{aligned}$$

If $f \colon A \longrightarrow A'$ is a morphism of K-algebras split by L, one sees immediately, via the isomorphisms above, that $\mathsf{Hom}_K(f, \mathsf{id}_L)$ is turned into $\mathsf{Sp}(\mathsf{id}_L \otimes f)$. $\square$

Chapter 9
The Galois Theorem for Rings

Convention. *In this chapter, all rings and algebras are commutative with a unit.*

Abstract The "Spectrum functor" of a ring and its right adjoint are the key to generalizing to the case of rings the notion of *split algebra* encountered in the case of fields. The same functors make it possible to define the profinite Galois groupoid of a Galois extension of rings. The Galois theorem for rings then exhibits an equivalence between the category of split algebras and that of profinite presheaves on the profinite Galois groupoid. In the case of fields, this reduces to the classical profinite Galois group and the Grothendieck Galois theorem for arbitrary Galois extensions of fields.

9.1 Split Algebras Over Rings

To ease the language, we split our approach into two definitions: being split over a ring, and being split by a ring homomorphism.

Definition 9.1 Let S be a ring. An S-algebra B is *split* when the counit

$$\alpha_B \colon C_S \mathsf{Sp}_S(B) \longrightarrow B$$

of the adjunction

$$\mathsf{Sp}_S \dashv C_S \colon \mathsf{Alg}_S^{\mathrm{op}} \underset{\longrightarrow}{\overset{\longleftarrow}{}} \mathsf{Prof}/\mathsf{Sp}(S)$$

(see Theorem 8.26) is an isomorphism.
We write Split_S for the category of split S-algebras.

Observe at once that:

Proposition 9.2 *Each ring S is a split S-algebra.*

Proof This is just a rephrasing of Theorem 8.23. $\qquad\qquad\square$

F. Borceux, *Galois Theories of Fields and Rings*, Coimbra Mathematical Texts 2,
https://doi.org/10.1007/978-3-031-58460-2_10

Definition 9.3 Let $\sigma\colon R \longrightarrow S$ be a ring homomorphism. An R-algebra A is split by σ when the S-algebra $S \otimes_R A$ is split.
We write $\mathsf{Split}_\sigma[S : R]$ for the category of R-algebras split by σ.

Our first example of such a situation is simply:

Proposition 9.4 *Let $\sigma\colon R \longrightarrow S$ be a ring homomorphism. The R-algebra R is split by σ.*

Proof We have $S \otimes_R R \cong S$ thus the result follows from Proposition 9.2. $\qquad\square$

We shall see later (Theorem 9.20) that Definition 9.3 generalizes Definition 2.23 of split algebra in the case of fields. But first, as suggested by Proposition 2.25, let us introduce further the following definition:

Definition 9.5 A ring homomorphism $\sigma\colon R \longrightarrow S$ is a *Galois morphism of rings* when the ring S, considered as an R-algebra, is split by σ.

Theorem 9.6 *Let $\sigma\colon R \longrightarrow S$ be a Galois morphism of rings. For each $(X, f) \in \mathsf{Prof/Sp}(S)$, the S-algebra $C_S(X, f)$, considered as an R-algebra, is split by σ.*

Proof For each $(X, f) \in \mathsf{Prof/Sp}(S)$, we have the following isomorphisms:

$$
\begin{aligned}
S \otimes_R C_S(X, f) &\cong S \otimes_R S \otimes_{C\left(\mathsf{Sp}(S),\mathbb{Z}\right)} C(X, \mathbb{Z}) \\
&\cong C_S \mathsf{Sp}_S(S \otimes_R S) \otimes_{C\left(\mathsf{Sp}(S),\mathbb{Z}\right)} C(X, \mathbb{Z}) \\
&\cong S \otimes_{C\left(\mathsf{Sp}(S),\mathbb{Z}\right)} C\left(\mathsf{Sp}(S \otimes_R S), \mathbb{Z}\right) \otimes_{C\left(\mathsf{Sp}(S),\mathbb{Z}\right)} C(X, \mathbb{Z}) \\
&\cong S \otimes_{C\left(\mathsf{Sp}(S),\mathbb{Z}\right)} C\left(\mathsf{Sp}(S \otimes_R S) \times_{\mathsf{Sp}(S)} X, \mathbb{Z}\right) \\
&\cong C_S\left(\mathsf{Sp}_S(S \otimes_R S) \times (X, f)\right) \\
&\cong C_S \mathsf{Sp}_S C_S\left(\mathsf{Sp}_S(S \otimes_R S) \times (X, f)\right) \\
&\cong C_S \mathsf{Sp}_S\left(S \otimes_R C_S(X, f)\right).
\end{aligned}
$$

These isomorphisms hold, successively,

- by Corollary 8.29 applied to (X, f);
- by assumption;
- by Corollary 8.29 applied to $\mathsf{Sp}_S(S \otimes_R S)$;
- by the form of pushouts in the category of rings (see Proposition 7.24) and because the contravariant functor

$$
C(-, \mathbb{Z})\colon \mathsf{Prof} \longrightarrow \mathsf{Rings},
$$

transforms pullbacks into pushouts (see Proposition 5.6; the functor $C(-, \mathbb{Z})$ has a right adjoint);
- again by Corollary 8.29;

- because the counit $\mathsf{Sp}_S C_S \Rightarrow \mathrm{id}$ of the adjunction $\mathsf{Sp}_S \dashv C_S$ is an isomorphism (see Theorem 8.26);
- by the first five isomorphisms.

The various naturalities guarantee that

$$S \otimes_R C_S(X, f) \cong C_S \mathsf{Sp}_S \big(S \otimes_R C_S(X, f)\big)$$

is the unit of the adjunction. Thus the S-algebra $S \otimes_R C_S(X, f)$ is split and therefore the R-algebra $C_S(X, f)$ is split by σ. □

Corollary 9.7 *Let S be a ring. The functors Sp_S and C_S induce a contravariant equivalence of categories*

$$\mathsf{Split}_S \underset{\mathsf{Sp}_S}{\overset{C_S}{\longleftrightarrow}} \mathsf{Prof}/\mathsf{Sp}(S).$$

Proof We have $S \otimes_S B \cong B$ for each S-algebra B, thus B is a split S-algebra precisely when B is split by the identity $S == S$. And by Proposition 9.2, the identity $S == S$ is a Galois morphism of rings. Moreover by Theorem 9.6, each S-algebra $C_S(X, f)$ is split. Thus the functor C_S takes values in the category of split S-algebras. We know already that the counit of the adjunction $\mathsf{Sp}_S \dashv C_S$ is an isomorphism (see Theorem 8.26). For each split algebra B, the unit $C_S \mathsf{Sp}_S(B) \cong B$ is an isomorphism as well, by definition of a split algebra. □

9.2 Galois Extensions of Rings

We are now ready to define the Galois extensions of rings.

Definition 9.8 A *Galois extension* of rings is a ring homomorphism $\sigma \colon R \longrightarrow S$ such that:

1. σ is a pure extension of rings (see Definition 7.18);
2. σ is a Galois morphism of rings (see Definition 9.5).

This section investigates the monadicity properties of a Galois extension of rings.

Lemma 9.9 *Let $\sigma \colon R \longrightarrow S$ be a Galois extension of rings. Every split S-algebra B is also an R-algebra split by σ; in particular $S \otimes_R B$ is a split S-algebra.*

Proof By assumption, $C_S \mathsf{Sp}_S(B) \cong B$. By Theorem 9.6, putting $(X, f) = \mathsf{Sp}_S(B)$, we obtain that the R-algebra B is split by σ. By Definition 9.3, this implies that the S-algebra $S \otimes_R B$ is split. □

Proposition 9.10 *Let $\sigma \colon R \rightarrowtail S$ be a Galois extension of rings. The functor*

$$S \otimes_R - \colon \mathsf{Split}_\sigma [S : R] \longrightarrow \mathsf{Split}_S$$

is co-monadic.

Proof An R-algebra A is split by σ when the S-algebra $S \otimes_R A$ is split (see Definition 9.3). This proves that the functor in the statement is correctly defined. To prove its monadicity, we shall use the Beck criterion (see Theorem 5.17)... or more precisely, its dual.

By Lemma 9.9, there exists a functor in the other direction

$$U_\sigma: \mathsf{Split}_S \longrightarrow \mathsf{Split}_\sigma[S:R], \quad B \mapsto B.$$

The two functors involved are simply the restrictions of the functors in Proposition 7.25, where U_σ is right adjoint to $S \otimes_R -$. The required natural bijections for having an adjunction (see Definition 5.1) thus hold in particular for the objects in the restricted categories.

By Theorem 7.27, the functor $S \otimes_R -$ at the level of all algebras is co-monadic, thus it reflects isomorphisms, by the Beck criterion. Thus its restriction to $\mathsf{Split}_\sigma[S:R]$ reflects isomorphisms as well.

Finally, consider two morphisms $u, v: A \Rightarrow A'$ in $\mathsf{Split}_\sigma[S:R]$ such that the morphisms $\mathsf{id}_S \otimes u$ and $\mathsf{id}_S \otimes v$ have a split equalizer in Split_S:

$$
K \xrightarrow[\ k\ ]{\ r\ } S \otimes_R A \underset{\mathsf{id} \otimes v}{\overset{\mathsf{id} \otimes u}{\rightrightarrows}} S \otimes_R A'.
$$

This split equalizer in Split_S is preserved by every functor (see Lemma 5.15) thus in particular by the inclusion in Alg_S. Theorem 7.27 and the Beck criterion in the case of the functor

$$S \otimes_R -: \mathsf{Alg}_R \longrightarrow \mathsf{Alg}_S$$

imply the existence in Alg_R of an equalizer

$$N \xrightarrow{\ n\ } A \underset{v}{\overset{u}{\rightrightarrows}} A'$$

preserved by $S \otimes_R -$. This means the existence of an isomorphism

$$\left(k: K \rightarrowtail S \otimes_R A\right) \cong \left(\mathsf{id}_S \otimes n: S \otimes_R N \rightarrowtail S \otimes_R A\right).$$

In particular, $S \otimes_R N \cong K \in \mathsf{Split}_S$; but this means that $N \in \mathsf{Split}_\sigma[S:R]$, by Definition 9.3. We thus have an equalizer $n = \mathrm{Ker}\,(u, v)$ in $\mathsf{Split}_\sigma[S:R]$ whose image under the functor $S \otimes_R -$ is the given split equalizer in Split_S. □

Theorem 9.11 Let $\sigma: R \longrightarrow S$ be a Galois extension of rings. The functor

$$\left(\mathsf{Split}_\sigma[S:R]\right)^{\mathrm{op}} \longrightarrow \mathsf{Prof}/\mathsf{Sp}(S), \quad A \mapsto \mathsf{Sp}_S(S \otimes_R A)$$

is co-monadic.

Proof This functor is the composite of the co-monadic functor of Proposition 9.10 and the contravariant equivalence of Corollary 9.7. □

9.3 The Profinite Galois Groupoid

This section defines the Galois groupoid of a Galois extension of rings, which will be a profinite groupoid. In the case of fields, one recaptures the corresponding profinite Galois group of Definition 4.5.

Lemma 9.12 *Let* $\sigma: R \longrightarrow S$ *be a Galois extension of rings. For each natural number* $0 \neq n \in \mathbb{N}$, *the* R-*algebra* $\otimes_{i=1}^{n} S = S \otimes_R \cdots \otimes_R S$ *is split by* σ.

Proof Let A be an R-algebra split by σ. This means that $S \otimes_R A$ is a split S-algebra (see Definition 9.3). Thus by Lemma 9.9, $S \otimes_R A$ is also an R-algebra split by σ. It suffices to start with $A = R$ (see Proposition 9.4) and iterate the argument.   □

Let us now put in a single diagram all the functors considered up to now in this specific problem.

$$\mathsf{Split}_\sigma[S:R]) \underset{S \otimes_R -}{\overset{\mathsf{U}_\sigma}{\underset{\longrightarrow}{\longleftarrow}}} \mathsf{Split}_S \underset{\mathsf{Sp}_S}{\overset{C_S}{\underset{\longrightarrow}{\longleftarrow}}} \mathsf{Prof}/\mathsf{Sp}(S)$$

$$\mathsf{Alg}_R \underset{S \otimes_R -}{\overset{\mathsf{U}_\sigma}{\underset{\longrightarrow}{\longleftarrow}}} \mathsf{Alg}_S \underset{\mathsf{Sp}_S}{\overset{C_S}{\underset{\longrightarrow}{\longleftarrow}}} \mathsf{Prof}/\mathsf{Sp}(S)$$

Theorem 9.13 *Let* $\sigma: R \longrightarrow S$ *be a Galois extension of rings. Consider the following diagram in the category of* R-*algebras*

$$(S \otimes_R S) \otimes_S (S \otimes_R S) \underset{\tau}{\overset{\varsigma_1 \otimes \varsigma_2}{\longleftarrow}} S \otimes_R S \overset{\varsigma_1}{\underset{\varsigma_2}{\overset{\nu}{\rightleftarrows}}} S$$

where

$$\varsigma_1(a) = a \otimes 1, \quad \varsigma_2(a) = 1 \otimes a, \quad \nu(a \otimes b) = ab, \quad \tau(a \otimes b) = b \otimes a.$$

The contravariant "spectrum" functor

$$\mathsf{Sp}: \mathsf{Alg}_R \longrightarrow \mathsf{Prof}$$

transforms this diagram into a profinite groupoid $\mathsf{Gal}_\sigma[S:R]$.

Proof The statement thus means that the following elements constitute a profinite groupoid $\mathcal{G}$ (see Section 5.5):

- the objects of $\mathcal{G}$ are the elements of $\mathsf{Sp}(S)$;
- the morphisms of $\mathcal{G}$ are the elements of $\mathsf{Sp}(S \otimes_R S)$, with $\mathsf{Sp}(\varsigma_1)$ and $\mathsf{Sp}(\varsigma_2)$ the mappings "domain" and "codomain"; that is, given $X, Y \in \mathsf{Sp}(S)$,

$$\mathcal{G}(X,Y) = \{f \in \mathsf{Sp}(S \otimes_R S) \big| \mathsf{Sp}(\varsigma_1)(f) = X, \ \mathsf{Sp}(\varsigma_2)(f) = Y\};$$

- for each object $X \in \mathsf{Sp}(S)$, the morphism $\mathsf{Sp}(v)(X)$ is the identity id_X;
- the mapping $\mathsf{Sp}(\varsigma_1 \otimes \varsigma_2)$ defines the composition of morphisms;
- the mapping $\mathsf{Sp}(\tau)$ defines the inverse operation.

The major difficulty in proving that we have a profinite groupoid concerns the "composition" morphism $\mathsf{Sp}(\varsigma_1 \otimes \varsigma_2)$: it must be defined on the space of pairs of composable morphisms, that is, the diagram

$$
\begin{array}{ccc}
\mathsf{Sp}\big((S \otimes_R S) \otimes_S (S \otimes_R S)\big) & \longrightarrow & \mathsf{Sp}(S \otimes_R S) \\
\big\downarrow & & \big\downarrow {\scriptstyle \mathsf{Sp}(\varsigma_2)} \\
\mathsf{Sp}(S \otimes_R S) & \xrightarrow[\mathsf{Sp}(\varsigma_1)]{} & \mathsf{Sp}(S)
\end{array}
$$

must be a pullback of profinite spaces.

To prove this fact, observe first that we have a pushout in the category of S-modules

$$
\begin{array}{ccc}
S & \xrightarrow{\varsigma_1} & S \otimes_R^1 S \\
{\scriptstyle \varsigma_2}\big\downarrow & & \big\downarrow {\scriptstyle t_2} \\
S \otimes_R^2 S & \xrightarrow[t_1]{} & (S \otimes_R^2 S) \otimes_S (S \otimes_R^1 S)
\end{array}
$$

where $S \otimes_R^1 S$ is the R-module $S \otimes_R S$ with the S-module structure

$$a(b \otimes c) = (ab) \otimes c;$$

and in an analogous way, $S \otimes_R^2 S$ is the R-module $S \otimes_R S$ with the S-module structure

$$a(b \otimes c) = b \otimes (ac).$$

The morphism t_1 is

$$t_2(b \otimes c) = (1 \otimes 1) \otimes (b \otimes c)$$

and analogously for t_2. We have an isomorphism of R-algebras

$$(S \otimes_R^2 S) \otimes_S (S \otimes_R^1 S) \cong S \otimes_R S \otimes_R S$$

and via this isomorphism, the S-module structure of the pushout becomes

$$s \left(\sum_i u_i \otimes v_i \otimes w_i \right) = \sum_i u_i \otimes (sv_i) \otimes w_i.$$

But each object in this pushout diagram is a split S-algebra:

- S is split by Proposition 9.2;
- S is an R-algebra split by σ (Definition 9.5), thus $S \otimes_R^1 S$ is a split S-algebra (Definition 9.3);
- we have an isomorphism of S-algebras

$$S \otimes_R^1 S \longrightarrow S \otimes_R^2 S, \quad s \otimes s' \mapsto s' \otimes s,$$

  thus $S \otimes_R^2 S$ is split because $S \otimes_R^1 S$ is split;
- $S \otimes_R S$ is split by σ (Lemma 9.12), thus $S \otimes_R S \otimes_R S$ with the S-module structure on the first component is a split S-algebra (see Definition 9.3); again the pushout object is an S-algebra isomorphic to that split S-algebra, via the isomorphism which permutes the first two components.

Thus finally, we have a pushout in the category Split_S of split S-algebras. By Corollary 9.7, the functor Sp_S transforms this pushout into a pullback in the category $\mathsf{Prof}/\mathsf{Sp}(S)$. But pullbacks in the category $\mathsf{Prof}/\mathsf{Sp}(S)$ are computed as in Prof. Thus indeed, we have the expected pullback in the category of profinite spaces.

We thus know already that the functor Sp transforms the diagram of the statement into a diagram of profinite spaces

$$\mathsf{Sp}(S \otimes_R S) \times_{\mathsf{Sp}(S)} \mathsf{Sp}(S \otimes_R S) \xrightarrow[\mathsf{Sp}(\tau) \, \uparrow]{\mathsf{Sp}(\varsigma_1 \otimes \varsigma_2)} \mathsf{Sp}(S \otimes_R S) \xleftarrow{\mathsf{Sp}(\nu)} \underset{\xrightarrow{\mathsf{Sp}(\varsigma_2)}}{\overset{\xrightarrow{\mathsf{Sp}(\varsigma_1)}}{}} \mathsf{Sp}(S)$$

where the left-hand object is the pullback of $\mathsf{Sp}(\varsigma_1)$ and $\mathsf{Sp}(\varsigma_2)$. It remains to check the axioms of a groupoid.

If $X \xrightarrow{f} Y \xrightarrow{g} Z$ is a pair of composable morphisms, we must check that the composite $\mathsf{Sp}(\varsigma_1 \otimes \varsigma_2)(f, g)$ is a morphism $X \longrightarrow Z$. The case of the domain means

$$\mathsf{Sp}(\varsigma_1)\big(\mathsf{Sp}(t_1)(f, g)\big) = \mathsf{Sp}(\varsigma_1)\big(\mathsf{Sp}(\varsigma_1 \otimes \varsigma_2)(f, g)\big).$$

It suffices to verify that

$$(\varsigma_1 \otimes \varsigma_2) \circ \varsigma_1 = t_1 \circ \varsigma_1$$

which is the case because:

$$\big((\varsigma_1 \otimes \varsigma_2) \circ \varsigma_1\big)(a) = (a \otimes 1) \otimes (1 \otimes 1) = (t_1 \circ \varsigma_1)(a).$$

The case of the codomain is analogous.

If X is an object in $\mathsf{Gal}_\sigma[S : R]$, we must first prove that $\mathsf{Sp}(\nu)(X)$ is a morphism $X \longrightarrow X$. That means

$$\big(\mathsf{Sp}(\varsigma_1) \circ \mathsf{Sp}(\nu)\big)(X) = X = \big(\mathsf{Sp}(\varsigma_2) \circ \mathsf{Sp}(\nu)\big)(X).$$

It suffices to prove that

$$\nu \circ \varsigma_1 = \mathsf{id}_S = \nu \circ \varsigma_2$$

and this holds because

$$(v \circ \varsigma_1)(a) = a1 = a = 1a = (v \circ \varsigma_2)(a).$$

We must next prove that for each morphism $f: X \longrightarrow Y$ in $\mathrm{Gal}_\sigma[S:R]$, $f \circ \mathrm{id}_X = f = \mathrm{id}_Y \circ f$. The first equality means

$$\mathrm{Sp}(\varsigma_1 \otimes \varsigma_2)\Big((\mathrm{Sp}(v) \circ \mathrm{Spec}(\varsigma_1))(f), f\Big) = f.$$

It suffices to prove that

$$\big((\varsigma_1 \circ v), \mathrm{id}\big) \circ (\varsigma_1 \otimes \varsigma_2)\big) = \mathrm{id}_{S \otimes_R S}.$$

This is true because

$$\begin{aligned}
\big((\varsigma_1 \circ v, \mathrm{id}) \circ (\varsigma_1 \otimes \varsigma_2)\big)(a \otimes b) &= (\varsigma_1 \circ v, \mathrm{id})\big((a \otimes 1) \otimes (1 \otimes b)\big) \\
&= (a \otimes 1) \cdot (1 \otimes b) \\
&= a \otimes b.
\end{aligned}$$

The proof of the second affirmation is analogous.

Now if $f: X \longrightarrow Y$ is a morphism, $f^{-1} = \mathrm{Sp}(\tau)(f)$ must be a morphism $f^{-1}: Y \longrightarrow X$. This means

$$(\mathrm{Sp}(\varsigma_1) \circ \tau)(f) = \mathrm{Sp}(\varsigma_2)(f), \quad (\mathrm{Sp}(\varsigma_2) \circ \tau)(f) = \mathrm{Sp}(\varsigma_1)(f).$$

It suffices to check that

$$\tau \circ \varsigma_1 = \varsigma_2, \quad \tau \circ \varsigma_2 = \varsigma_1.$$

This is the case since

$$(\tau \circ \varsigma_1)(a) = \tau(a \otimes 1) = 1 \otimes a = \varsigma_2(a)$$

and in the same way for the second equality.

The equality $f \circ f^{-1} = \mathrm{id}_Y$ means

$$\mathrm{Sp}(\varsigma_1 \otimes \varsigma_2)\big(\mathrm{Sp}(\tau)(f), f\big) = \big(\mathrm{Sp}(v) \circ \mathrm{Sp}(\varsigma_2)\big)(f).$$

It suffices to prove that

$$(\tau, \mathrm{id}) \circ (\varsigma_1 \otimes \varsigma_2) = \varsigma_2 \circ v.$$

This is true because:

$$\big((\tau, \mathrm{id}) \circ (\varsigma_1 \otimes \varsigma_2)\big)(a, b) = (1 \otimes a) \cdot (1 \otimes b) = 1 \otimes ab = (\varsigma_2 \circ v)(a \otimes b).$$

The proof of the equality $f^{-1} \circ f = \mathrm{id}_X$ is analogous.

Finally we must prove the associativity of the composition. As above, the pushout of R-algebras

$$
\begin{array}{ccc}
S \otimes_R S & \xrightarrow{t_1} & (S \otimes_R S) \otimes_S (S \otimes_R S) \\
\downarrow{\scriptstyle t_2} & & \downarrow{\scriptstyle r_2} \\
(S \otimes_R S) \otimes_S (S \otimes_R S) & \xrightarrow{r_1} & (S \otimes_R S) \otimes_S (S \otimes_R S) \otimes_S (S \otimes_R S)
\end{array}
$$

is also a pushout of split S-algebras and this implies that

$$
\mathsf{Sp}\big((S \otimes_R S) \otimes_S (S \otimes_R S) \otimes_S (S \otimes_R S)\big) \\
\cong \mathsf{Sp}(S \otimes_R S) \times_{\mathsf{Sp}(S)} \mathsf{Sp}(S \otimes_R S) \times_{\mathsf{Sp}(S)} \mathsf{Sp}(S \otimes_R S)
$$

is the space of triples,

$$
W \xrightarrow{f} X \xrightarrow{g} Y \xrightarrow{h} Z
$$

of composable morphisms. The equality $(h \circ g) \circ f = h \circ (g \circ f)$ means

$$
\mathsf{Sp}(\varsigma_1 \otimes \varsigma_2)\big((\mathsf{Sp}(\varsigma_1 \otimes \varsigma_2)(h, g), f\big) = \mathsf{Sp}(\varsigma_1 \otimes \varsigma_2)\big(h, \mathsf{Sp}(\varsigma_1 \otimes \varsigma_2)(g, f)\big).
$$

It suffices to prove

$$
(\varsigma_1 \otimes \varsigma_2, \mathsf{id}) \circ (\varsigma_1 \otimes \varsigma_2) = (\mathsf{id}, \varsigma_1 \otimes \varsigma_2) \circ (\varsigma_1 \otimes \varsigma_2).
$$

Again this result holds because

$$
\big((\varsigma_1 \otimes \varsigma_2, \mathsf{id}) \circ (\varsigma_1 \otimes \varsigma_2)\big)(a \otimes b) = (a \otimes 1) \otimes (1 \otimes 1) \otimes (1 \otimes b) \\
= \big((\mathsf{id}, \varsigma_1 \otimes \varsigma_2) \circ (\varsigma_1 \otimes \varsigma_2)\big)(a \otimes b).
$$

This concludes the proof. $\qquad\square$

Let us mention that various aspects of the proof of Theorem 9.13 are special instances of a more general theorem: if $\sigma : R \longrightarrow S$ is a morphism in a category C with finite colimits, the cokernel pair $(\varsigma_1, \varsigma_2)$ of σ (the pushout of σ with itself) is an internal groupoid.

We can now finally define:

Definition 9.14 Let $\sigma : R \longrightarrow S$ be a Galois extension of rings. The profinite Galois groupoid $\mathsf{Gal}_\sigma[S : R]$ of this extension is the profinite groupoid

$$
\mathsf{Sp}(S \otimes_R S) \times_{\mathsf{Sp}(S)} \mathsf{Sp}(S \otimes_R S) \xrightarrow{\mathsf{Sp}(\varsigma_1 \otimes \varsigma_2)} \mathsf{Sp}(S \otimes_R S) \underset{\xrightarrow{\mathsf{Sp}(\varsigma_2)}}{\overset{\mathsf{Sp}(\varsigma_1)}{\underset{\xleftarrow{\mathsf{Sp}(v)}}{\rightrightarrows}}} \mathsf{Sp}(S)
$$

$\mathsf{Sp}(\tau)$

of Theorem 9.13.

9.4 The Galois Theorem for Rings

This section is devoted to the proof of the main theorem of this book:

Theorem 9.15 (Galois Theorem) *Let* $\sigma \colon R \longrightarrow S$ *be a Galois extension of rings and* $\mathsf{Gal}_\sigma[S : R]$ *its profinite Galois groupoid (see Definition 9.14). There exists an equivalence of categories*

$$\mathsf{Split}_\sigma[S : R] \approx \mathsf{Gal}_\sigma[S : R]\text{-Prof}$$

between the category of R-algebras split by σ *and the category of profinite presheaves on the profinite groupoid* $\mathsf{Gal}_\sigma[S : R]$.

Proof The category $\mathsf{Gal}_\sigma[S : R]$-Prof is the category of algebras for the monad on $\mathsf{Prof}/\mathsf{Sp}(S)$ described in Theorem 6.14. By Theorem 9.11, the category $\mathsf{Split}_\sigma[S : R]$ is also monadic over $\mathsf{Prof}/\mathsf{Sp}(S)$. To prove the Galois theorem, it suffices thus to prove that these two monads coincide.

The adjoint of the functor

$$\mathsf{Split}_\sigma[S : R] \longrightarrow \mathsf{Prof}/\mathsf{Sp}(S), \quad A \mapsto \mathsf{Sp}_S(S \otimes_R A)$$

is the functor

$$\mathsf{Prof}/\mathsf{Sp}(S) \xrightarrow{\quad C_S \quad} \mathsf{Split}_S \xrightarrow{\quad U_\sigma \quad} \mathsf{Split}_\sigma[S : R], \quad (X, f) \mapsto C_S(X, f)$$

(see Theorem 9.11, Corollary 9.7 and Proposition 9.10). The corresponding monad $\mathbb{T} = (T, \varepsilon, \mu)$ is such that

$$T \colon \mathsf{Prof}/\mathsf{Sp}(S) \longrightarrow \mathsf{Prof}/\mathsf{Sp}(S), \quad (X, f) \mapsto \mathsf{Sp}_S\big(S \otimes_R C_S(X, f)\big)$$

(see Proposition 5.8). First, let us prove that this functor T is isomorphic to the functor T of Theorem 6.14.

Consider $(X, f) \in \mathsf{Prof}/\mathsf{Sp}(S)$. With the notation of Theorem 9.13, consider the pushout of S-algebras

$$
\begin{array}{ccc}
S & \xrightarrow{\ s_1\ } & S \otimes_R^1 S \\[4pt]
{\scriptstyle \alpha_{(X,f)}}\Big\downarrow & & \Big\downarrow{\scriptstyle \delta_{(X,f)} \,\cong\, (\mathsf{id}_S \otimes \alpha_{(X,f)}) \circ \tau} \\[4pt]
C_S(X, f) & \xrightarrow[\ \gamma_{(X,f)}\ ]{} & (S \otimes_R^1 S) \otimes_S C_S(X, f) \cong S \otimes_R C_S(X, f)
\end{array}
$$

where

- $\delta_{(X,f)} \colon S \otimes_R^1 S \longrightarrow (S \otimes_R^1 S) \otimes C_S(X, f), \quad s \otimes s' \mapsto (s \otimes s') \otimes 1$;
- $\gamma_{(X,f)} \colon C_S(X, f) \longrightarrow S \otimes_R C_S(X, f), \quad a \mapsto 1 \otimes a$;
- $\alpha_{(X,f)} \colon S \longrightarrow C_S(X, f), \quad s \mapsto s \cdot 1$ with 1 the unit of the S-algebra (X, f);

- the S-module structure on $S \otimes_R C_S(X, f)$ is the action of S on $C_S(X, f)$, that is

$$S \otimes_R S \otimes_R C_S(X, f) \longrightarrow S \otimes_R C_S(X, f), \quad s \otimes s' \otimes a \mapsto s' \otimes \alpha_{(X,f)}(s) \cdot a;$$

- the mentioned isomorphism is

$$(S \otimes_R^1 S) \otimes_S C_S(X, f) \longrightarrow S \otimes_R C_S(X, f), \quad s \otimes s' \otimes a \mapsto s' \otimes sa;$$

- via this isomorphism, we have

$$s \otimes s' \otimes a = s(1 \otimes s') \otimes a = (1 \otimes s') \otimes sa \cong s' \otimes sa$$

and this implies in particular

$$\begin{aligned}
\delta_{(X,f)}(s \otimes s') &= (s \otimes s') \otimes 1 \\
&\cong s' \otimes s1 \\
&= (\mathsf{id}_S \otimes \alpha_{(X,f)})(s' \otimes s) \\
&= ((\mathsf{id}_S \otimes \alpha_{(X,f)}) \circ \tau)(s \otimes s').
\end{aligned}$$

Again previous results (Proposition 9.2, Theorem 9.6 and Lemma 9.9) imply that the four objects in the diagram are split S-algebras. Therefore the image of that pushout under the contravariant equivalence of Corollary 9.7 is a pullback in $\mathsf{Prof}/\mathsf{Sp}(S)$:

$$
\begin{array}{ccc}
\mathsf{Sp}_S(S) & \xleftarrow{\ \mathsf{Sp}(\varsigma_1)\ } & \mathsf{Sp}_S(S \otimes_R S) \\[4pt]
{\scriptstyle \mathsf{Sp}(\alpha_{(X,f)})} \big\uparrow & & \big\uparrow {\scriptstyle \mathsf{Sp}\left((\mathsf{id}_S \otimes \alpha_{(X,f)}) \circ \tau\right)} \\[4pt]
(X, f) \cong \mathsf{Sp}_S C_S(X, f) & \xleftarrow[\ \mathsf{Sp}(\gamma_{(X,f)})\ ]{} & \mathsf{Sp}_S(S \otimes_R C_S(X, f))
\end{array}
$$

with $\mathsf{Sp}_S C_S(X, f) \cong (X, f)$ by Theorem 8.26. But pullbacks in $\mathsf{Prof}/\mathsf{Sp}(S)$ are computed as in Prof. Thus we obtain a diagram in Prof, with the square a pullback.

$$
\begin{array}{ccc}
\mathsf{Sp}(S \otimes_R C_S(X, f)) & \xrightarrow{\ \mathsf{Sp}(\gamma_{(X,f)})\ } & X \\[4pt]
{\scriptstyle \mathsf{Sp}\left((\mathsf{id}_S \otimes \alpha_{(X,f)}) \circ \tau\right)} \big\downarrow & & \big\downarrow {\scriptstyle f} \\[4pt]
\mathsf{Sp}(S \otimes_R S) & \xrightarrow[\ \mathsf{Sp}(\varsigma_1)\ ]{} & \mathsf{Sp}(S) \\[4pt]
{\scriptstyle \mathsf{Sp}(\varsigma_2)} \big\downarrow & & \\[4pt]
\mathsf{Sp}(S) & &
\end{array}
$$

The image of (X, f) under the monad of Theorem 6.14 is the composite of the pullback of f with $\mathsf{Sp}(\varsigma_1)$ and the morphism $\mathsf{Sp}(\varsigma_2)$; that is, the morphism $\mathsf{Sp}(\varsigma_2) \circ \mathsf{Sp}\left(\left(\mathsf{id}_S \otimes \alpha_{(X,f)}\right) \circ \tau\right)$. The image of (X, f) under the monad of Theorem 9.11 is $\mathsf{Sp}_S\left(S \otimes_R C_S(X, f)\right)$, with the S-module structure

$$S \otimes_R S \otimes_R C_S(X, f) \longrightarrow S \otimes_R C_S(X, f), \quad s \otimes s' \otimes a \mapsto ss' \otimes a.$$

Now

$$\mathsf{Sp}_S C_S(X, f) = \left(\mathsf{Sp}\left(S \otimes_R C_S(X, f)\right), \mathsf{Sp}\left(\varphi_{(X,f)}\right)\right),$$

where

$$\varphi_{(X,f)} \colon S \longrightarrow S \otimes_R C_S(X, f), \quad s \mapsto s \otimes 1.$$

Observe that

$$\left(\left(\mathsf{id}_S \otimes \alpha_{(X,f)}\right) \circ \tau \circ \varsigma_2\right)(s) = s \otimes 1 = \varphi_{(X,f)}(s).$$

This proves that

$$\mathsf{Sp}_S\left(S \otimes_R C_S(X, f)\right) = \left(\mathsf{Sp}\left(S \otimes_R C_S(X, f)\right), \mathsf{Sp}(\varsigma_2) \circ \mathsf{Sp}\left(\left(\mathsf{id}_S \otimes \alpha_{(X,f)}\right) \circ \tau\right)\right);$$

that is, both monads coincide on the objects. It is straightforward to extend this conclusion to the case of morphisms.

We must next verify that both monads have the same unit and the same multiplication. The unit of the monad in Theorem 6.14 is, given an object $(X, f) \in \mathsf{Prof}/\mathsf{Sp}(S)$,

$$X \longrightarrow \mathsf{Sp}_S(S \otimes_R S) \times_{\mathsf{Sp}_S} X, \quad x \mapsto \left(\mathsf{id}_{f(x)}, x\right) = \left(\left(\mathsf{Sp}(\nu) \circ f\right)(x), x\right).$$

The unit of the monad in Theorem 9.11

$$(X, f) \longrightarrow \mathsf{Sp}_S\left(S \otimes_R C_S(X, f)\right)$$

corresponds by the contravariant equivalence of Corollary 9.7 to the counit

$$C_S(X, f) \longleftarrow S \otimes_R C_S(X, f)$$

of the comonad of Proposition 9.10. But the situation of Proposition 9.10 is the restriction of the situation of Proposition 7.25. This implies that the counit is the multiplication

$$m_{(X,f)} \colon S \otimes_R C_S(X, f) \longrightarrow C_S(X, f), \quad s \otimes a \mapsto sa.$$

To prove the equality of the two units in the category Prof, it suffices to prove the equality of the composites with both projections of the pullback. Via the isomorphism $\mathsf{Sp}_S C_S(X, f) \cong (X, f)$ of Theorem 8.26, this means

$$\mathsf{Sp}\Big(\big(\mathsf{id}_S \circ \alpha_{(X,f)}\big) \circ \tau\Big) \circ \mathsf{Sp}(m_{(X,f)}) \cong \mathsf{Sp}(v) \circ f = \mathsf{Sp}(v) \circ \mathsf{Sp}(C_S(f))$$

$$\mathsf{Sp}(\gamma_{(X,f)}) \circ \mathsf{Sp}(m_{(X,f)}) \cong \mathsf{id}_X = \mathsf{Sp}(\mathsf{id}_{C_S(X,f)}).$$

This is indeed the case because

$$C_S(f) \colon S \cong C_S\mathsf{Sp}_S(S) \longrightarrow C_S(X,f), \quad s \mapsto s1$$

and therefore

$$\Big(m_{(X,f)} \circ \big(\mathsf{id}_S \circ \alpha_{(X,f)}\big) \circ \tau\Big)(s \otimes s') = s's1 = ss'1 = C_S(f)\big(v(s \otimes s')\big)$$

$$\big(m_{(X,f)} \circ \gamma_{(X,f)}\big)(a) = 1a = a = \mathsf{id}_{C_S(X,f)}(a).$$

The proof of the equality of the multiplications is analogous. Again, since the situation of Proposition 9.10 is the restriction of the situation in Proposition 7.25, the comultiplication of the comonad in Proposition 9.10 is, given an object $C_S(X, f)$,

$$\theta_{(X,f)} \colon S \otimes_R C_S(X, f) \longrightarrow S \otimes_R S \otimes_R C_S(X, f), \quad s \otimes a \mapsto s \otimes 1 \otimes a.$$

We must prove that $\mathsf{Sp}(\theta_{(X,f)})$ is the multiplication of the monad in Theorem 6.14:

$$\mathsf{Sp}(S \otimes_R S) \times_{\mathsf{Sp}(S)} \mathsf{Sp}(S \otimes_R S) \times_{\mathsf{Sp}(S)} X \longrightarrow \mathsf{Sp}(S \otimes_R S) \times_{\mathsf{Sp}(S)} X$$

$$(g, h, x) \mapsto (h \circ g, x).$$

Again it suffices to prove the equality of the composites with both projections of the pullback above. Observe that

$$\mathsf{Sp}(S \otimes_R S) \times_{\mathsf{Sp}(S)} X \cong \mathsf{Sp}\big(S \otimes_R C_S(X, f)\big)$$

and a double application of this formula yields

$$\mathsf{Sp}(S \otimes_R S) \times_{\mathsf{Sp}(S)} \mathsf{Sp}(S \otimes_R S) \times_{\mathsf{Sp}(S)} X \cong \mathsf{Sp}\big(S \otimes_R S \otimes_R C_S(X, f)\big).$$

First,

$$\big(\theta_{(X,f)} \circ \gamma_{(X,f)}\big)(a) = 1 \otimes 1 \otimes a.$$

This proves that the morphism

$$\theta_{(X,f)} \circ \gamma_{(X,f)} \colon C_S(X, f) \longrightarrow (S \otimes_R S) \otimes_R C_S(X, f)$$

is the second injection of the pushout; thus, its image under the functor Sp is the second projection

$$(g, h, x) \mapsto x$$

of the pullback in the category Prof.

Second,

$$\big(\mu_{(X,f)} \circ (\mathsf{id}_S \otimes \alpha_{(X,f)}) \circ \tau\big)(s \otimes s') = s' \otimes 1 \otimes s1.$$

Via the iterated isomorphism

$$(S \otimes_R^1 S) \otimes_S (S \otimes_R^1 S) \otimes_S C_S(X, f) \cong S \otimes_R S \otimes_R C_S(X, f)$$

and using the corresponding S-module structure of $S \otimes_R C_S(X, f)$ (see above) this element corresponds to the element

$$(s \otimes 1) \otimes (1 \otimes s') \otimes 1 \in (S \otimes_R^1 S) \otimes_S (S \otimes_R^1 S) \otimes_S C_S(X, f).$$

This is precisely the image of the element

$$(\varsigma_1 \otimes \varsigma_2) \otimes (s \otimes s') \in (S \otimes_R^1 S) \otimes_S (S \otimes_R^1 S)$$

under the injection

$$(S \otimes_R^1 S) \otimes_S (S \otimes_R^1 S) \longrightarrow (S \otimes_R^1 S) \otimes_S (S \otimes_R^1 S) \otimes_S C_S(X, f)$$

of the pushout. It is therefore the image under the functor Sp of the mapping

$$(g, h, x) \mapsto h \circ g. \qquad \qquad \square$$

9.5 The Case of Fields

To conclude this book, it remains to observe that the Galois Theorem for rings (Theorem 9.15) contains as a particular case the most general Galois Theorem for fields (Theorem 4.24).

Lemma 9.16 *Let $\sigma \colon K \rightarrowtail L$ be a finite-dimensional extension of fields. Each finite-dimensional K-algebra A split by L in the sense of Definition 2.23, is also split by σ in the sense of Definition 9.3.*

Proof The spectrum of a field is a singleton (see Proposition 8.31). This implies

$$\mathsf{Prof}/\mathsf{Sp}(L) \cong \mathsf{Prof}/\{*\} \cong \mathsf{Prof}.$$

The structural space of L is the set L with the discrete topology. The functor Sp_L is simply

$$\mathsf{Sp}_L = \mathsf{Sp} \colon \mathsf{Alg}_L \longrightarrow \mathsf{Prof}.$$

The functor C_L is

$$C_L = C(-, L) \colon \mathsf{Prof} \longrightarrow \mathsf{Alg}_L, \quad X \mapsto C(X, L)$$

where L has the discrete topology.

Let A be a finite-dimensional K-algebra split by L in the sense of Definition 2.23. We must prove that the canonical morphism

$$C\big(\mathsf{Sp}(L \otimes_K A), L\big) \longrightarrow L \otimes A$$

is an isomorphism. But we have an isomorphism

$$\mathsf{Gel}_A : L \otimes_K A \cong L^n$$

where $n = \dim{}_K A$ (see Theorem 2.27). Moreover, the functor $\mathsf{Sp} \colon \mathsf{Alg}_R \longrightarrow \mathsf{Prof}$ transforms limits into colimits, because it has an adjoint (see Theorem 8.26). This implies

$$C\big(\mathsf{Sp}(L \otimes_K A), L\big) \cong C\big(\mathsf{Sp}(L^n), L\big) \cong C\big(n, L\big) \cong L^n \cong L \otimes_K A,$$

where

$$n = \amalg_{i=1}^n \{*\} = \amalg_{i=1}^n \mathsf{Sp}(L) = \mathsf{Sp}(L^n)$$

is the discrete space with n elements. $\qquad\square$

Corollary 9.17 *Each finite-dimensional Galois extension $\sigma \colon K \rightarrowtail L$ of fields (see Definition 1.12) is also a Galois extension of rings (see Definition 9.8).*

Proof By Proposition 7.19, Lemma 9.16 and Proposition 2.25. $\qquad\square$

Lemma 9.18 *Let $\sigma \colon K \rightarrowtail L$ be an arbitrary extension of fields. Each K-algebra A split by L in the sense of Definition 2.23, is also split by σ, in the sense of Definition 9.3.*

Proof As in Lemma 9.16, if A is a K-algebra split by L in the sense of Definition 2.23, we must prove that the canonical morphism

$$C\big(\mathsf{Sp}(L \otimes_K A), L\big) \longrightarrow L \otimes A$$

is an isomorphism.

The algebra A is the filtered union of its finite-dimensional K-subalgebras $B \subseteq A$ (see Proposition 4.20). Moreover, by Proposition 4.21, each finite-dimensional subalgebra B is split, in the sense of Definition 2.23, by some finite-dimensional subextension $K \subseteq M_B \subseteq L$. Those filtered unions are special cases of filtered colimits and the "tensor product functor" preserves colimits, because it has a right adjoint functor (see Example 5.2 and Proposition 2.10). We get then

$$L \otimes_K A \cong \Big(\operatorname*{colim}_M M\Big) \otimes_K \Big(\operatorname*{colim}_B B\Big) \cong \operatorname*{colim}_{(M,B)} M \otimes_K B.$$

Observe that when B is split by M, trivially B is split by every Galois extension $M' \supseteq M$. Since the finite-dimensional Galois subextensions $K \subseteq M \subseteq L$ constitute a filtered family (see Proposition 4.3), the pairs (M, B) such that B is split by M

constitute a cofinal[1] part in the partially ordered set of indices (M, B). Thus

$$L \otimes_K A \cong \operatorname*{colim}_{(M,B)} M \otimes_K B \quad \text{with } B \text{ split by } M.$$

The functor $\mathsf{Sp} \colon \mathsf{Alg}_R \longrightarrow \mathsf{Prof}$ transforms filtered colimits into co-filtered limits (see Proposition 8.5). The functor $C(-, L) \colon \mathsf{Prof} \longrightarrow \mathsf{Alg}_L$ transforms limits into colimits, because it has an adjoint (see Theorem 8.26). Considering only the indices (M, B) where B and M are finite-dimensional and B is split by M, we get

$$C\big(\mathsf{Sp}(M \otimes_K B), M\big) \cong M \otimes_K B$$

by Lemma 9.16. This implies

$$
\begin{aligned}
C\big(\mathsf{Sp}(L \otimes_K A), L\big) &\cong C\big(\mathsf{Sp}(\operatorname{colim}_{(M,B)} M \otimes_K B), L\big) \\
&\cong C\big(\lim_{(M,B)} \mathsf{Sp}(M \otimes_K B), L\big) \\
&\cong \operatorname{colim}_{(M,B)} C\big(\mathsf{Sp}(M \otimes_K B), L\big) \\
&\cong \operatorname{colim}_{(M,B)} C\big(\mathsf{Sp}(M \otimes_K B), \operatorname{colim}_{M'} M'\big) \\
(1) \quad &\cong \operatorname{colim}_{(M,B,M')} C\big(\mathsf{Sp}(M \otimes_K B), M'\big) \\
(2) \quad &\cong \operatorname{colim}_{(M,B)} C\big(\mathsf{Sp}(M \otimes_K B), M\big) \\
&\cong \operatorname{colim}_{(M,B)} M \otimes_K B \\
&\cong L \otimes_K A.
\end{aligned}
$$

The isomorphism (1) holds because $\mathsf{Sp}(M \otimes_K B)$ is a finite discrete space (see Proposition 8.35). If

$$f \colon \{x_1, \dots, x_m\} = \mathsf{Sp}(M \otimes_K B) \longrightarrow \operatorname*{colim}_{M'} M'$$

is a (necessarily continuous) mapping, each $f(x_i)$ belongs to some extension M_i'. But the extensions M' constitute a filtered family (see Proposition 4.3), thus there exists some M' containing each M_i'. So f factors through M'. The same Proposition 4.3 also implies the isomorphism (2). For each triple (M, B, M') we can choose M'' containing M and M'. A fortiori, B is split by M''. Thus the triples (M'', B, M'') with B split by M'' constitute a cofinal part in the partially ordered set of the triples of indices and it suffices therefore to compute the colimit on theses indices.   □

Corollary 9.19 *Every Galois extension* $\sigma \colon K \rightarrowtail L$ *of fields (see Definition 1.12) is also a Galois extension of rings (see Definition 9.8).*

Proof By Proposition 7.19, Lemma 9.18 and Proposition 2.25.   □

Finally, we can conclude that:

---

[1] A subset $Y \subseteq X$ of a partially ordered set $(X, \leq)$ is *cofinal* when for each element $x \in X$ there exists an element $y \in Y$ such that $x \leq y$.

Theorem 9.20 *Let* $\sigma: K \rightarrowtail L$ *be a Galois extension of fields, in the sense of Definition 1.12. The Galois Theorem for rings (see Theorem 9.15) reduces to the Grothendieck Galois Theorem for fields, in arbitrary dimension (see Theorem 4.24).*

Proof By Corollary 9.19, σ is a Galois extension in the sense of Definition 9.8.

The Galois groupoid is a profinite group G, because $\mathsf{Sp}(L)$ is a singleton (see Proposition 8.31 and Proposition 6.4). This group has as set of elements

$$G = \mathsf{Sp}(L \otimes_K L) \cong \mathsf{Hom}_K(L, L) = \mathsf{Gal}[L : K]$$

by Proposition 8.35.

We must now prove that the multiplication of G corresponds to the composition of K-automorphisms of L. The K-algebra L is split by σ (Definition 9.5 and Corollary 9.19). Thus the equivalence

$$\mathsf{Split}_\sigma[L : K] \cong \mathsf{Gal}_\sigma[L : K]\text{-Prof} \cong G\text{-Prof}$$

induces a contravariant isomorphism between groups of automorphisms

$$\mathsf{Hom}_K(L, L) \cong \mathsf{Hom}_G\big(\mathsf{Sp}(L \otimes_K L), \mathsf{Sp}(L \otimes_K L)\big) = \mathsf{Hom}_G(G, G),$$

where $G = \mathsf{Sp}(L \otimes_K L)$ is considered as a profinite G-space. It thus remains to prove the existence of a contravariant group isomorphism $G \cong \mathsf{Hom}_G(G, G)$.

Each endomorphism $\psi: G \longrightarrow G$ necessarily has the form

$$\psi: G \longrightarrow G, \quad x \mapsto \psi(x) = \psi(x1) = x\psi(1).$$

Conversely, each element $g \in G$ induces a continuous endomorphism

$$\varphi_g: G \longrightarrow G, \quad x \mapsto \varphi_g(x) = xg.$$

All these endomorphisms are automorphisms, with $\varphi_{g^{-1}}$ the inverse of φ_g. We see immediately that we have described a contravariant isomorphism of groups:

$$G \cong \mathsf{Hom}_G(G, G), \quad g \mapsto \varphi_g.$$

Proposition 6.13 implies that the category $\mathsf{Gal}[L : K]$-Prof of Theorem 4.24 coincides with the category $\mathsf{Gal}_\sigma[L : K]$-Prof of Theorem 9.15. This yields the equality in the following diagram.

$$
\begin{array}{ccc}
\mathsf{Split}[L : K] & \xrightarrow[\cong]{\mathsf{Hom}_K(-, L)} & \mathsf{Gal}[L : K]\text{-Prof} \\[2pt]
\downarrow & & \| \\[4pt]
\mathsf{Split}_\sigma[L : K] & \xrightarrow[\mathsf{Sp}(L \otimes_K -)]{\cong} & \mathsf{Gal}_\sigma[S : R]\text{-Prof}
\end{array}
$$

The upper line is an equivalence by Theorem 4.24 and the lower line is the equivalence of Theorem 9.15. The left-hand inclusion exists by Lemma 9.18. The diagram is commutative by Proposition 8.35. This implies that the left-hand inclusion is an equivalence of categories. □

Further Reading

1. **M. Auslander and O. Goldman**, The Brauer group of a commutative ring, *Trans. Amer. Math. Soc.* **97**, 1960, 367–409
2. **M. Barr**, Exact categories. In: M. Barr, P.A. Grillet and D.H. van Osdol *Exact Categories and Categories of Sheaves*. Springer Lect. Notes in Math. **236**, 1971, 1–120
3. **M. Barr**, Abstract Galois theory, *J. Pure and Appl. Algebra* **19**, 1980, 21–42
4. **M. Barr**, Abstract Galois theory II, *J. Pure Appl. Algebra* **25**, 1982, 227–247
5. **M. Barr and J. Beck**, Homology and standard constructions. In: Eckmann, B. (ed) *Seminar on Triples and Categorical Homology Theory*. Springer Lect. Notes in Math. **80** (1969) 245–335
6. **F. Borceux**, *Handbook of Categorical Algebra*, 3 volumes, Cambridge University Press, 1994
7. **F. Borceux and G. Janelidze**, *Galois Theories*, Cambridge University Press, 2001
8. **F. Borceux**, *Algumas teorias de Galois dos Corpos e dos Anéis*, Textos de Matemática **35**, Universidade de Coimbra, 2004
9. **N. Bourbaki**, *Algèbre*, chapitre IV, Masson, Paris, 1981
10. **N.Bourbaki**, *Topologie générale*, Ch. **2**, Hermann, Paris, 1971
11. **A. Carboni, G. Janelidze, and A.R. Magid**, A note on Galois correspondence for commutative rings, *J. Algebra* **183**, 1996, 266–272
12. **S.U. Chase, D.K. Harrison, and A. Rosenberg**, *Galois Theory and Cohomology of Commutative Rings*, Mem. Amer. Math. Soc. **52**, 1965
13. **S.U. Chase and M.E. Sweedler**, *Hopf Algebras and Galois Theory*, Springer Lect. Notes in Math. **97**, 1969
14. **F. DeMeyer and E. Ingraham**, *Separable Algebras over Commutative Rings*, Springer Lect. Notes in Math. **181**, 1971
15. **R. et A. Douady**, *Algèbre et théories galoisiennes*, Fernand Nathan, 1977
16. **A. Grothendieck**, *Revêtements étales et groupe fondamental*, SGA1, exposé V, Springer Lect. Notes in Math. **224**, 1971
17. **G. Janelidze**, Magid's theorem in categories, *Bull. Georgian Acad. Sci.* **114**, 3, 1984, 497–500 (in Russian)
18. **G. Janelidze**, The fundamental theorem of Galois theory, *Math. USSR Sbornik* **64** (2), 1989, 359–374
19. **G. Janelidze**, Pure Galois theory in categories, *J. Algebra* **132**, 1990, 270–286
20. **G. Janelidze**, What is a double central extension? (the question was asked by Ronald Brown), *Cahiers de Topologie et Géometrie Différentielle Catégorique* **XXXII-3**, 1991, 191–202
21. **G. Janelidze**, Precategories and Galois theory. In: Carboni, A., Pedicchio, M.C., Rosolini, G. (eds) *Category Theory*. Springer Lect. Notes in Math. **1488**, 1991, 157–173
22. **G. Janelidze and G.M. Kelly**, Galois theory and a general notion of central extension, *J. Pure Appl. Algebra* **97**, 1994, 135–161
23. **G. Janelidze, D. Schumacher, R. Street**, Galois theory in variable categories, *Appl. Categ. Structures* **1**, 1993, 103–110

F. Borceux, *Galois Theories of Fields and Rings*, Coimbra Mathematical Texts 2,
https://doi.org/10.1007/978-3-031-58460-2

24. **G. Janelidze and R.H. Street**, Galois theory in symmetric monoidal categories, *J. of Algebra 220*, 1999, 174-187

25. **G. Janelidze and W. Tholen**, Facets of descent, I, *Appl. Categ. Structures* **2**, 1994, 245–281

26. **G.J. Janusz**, Separable algebras over commutative rings, *Trans. Amer. Math. Soc.* **122**, 1966, 461–479

27. **A. Joyal and M. Tierney**, *An Extension of the Galois Theory of Grothendieck*, Mem. Amer. Math. Soc. **51**, 309, 1984

28. **Kelley**, *General Topology*, Springer, 1971 (from Van Nostrand, 1955)

29. **Th.S. Ligon**, *Galois-Theorie in monoidalen Kategorien*, Algebra Berichten, 1978

30. **S. Mac Lane**, *Categories for the Working Mathematician*, Second Ed., Springer, 1998

31. **A.R. Magid**, *The Separable Galois Theory of Commutative Rings*, Marcel Dekker, 1974

32. **B. Mesablishvili**, Pure morphisms of commutative rings are effective descent morphisms for modules – A new proof, *Th. and Appl. of Categ.*, 7-3, 2000, 38–42

33. **J.P. Olivier**, Descente par morphismes purs, *C. R. Acad. Sc. Paris*, **271-A**, 1970, 821–823

34. **I. Stewart**, *Galois Theory*, Chapman and Hall, 1973

35. **O.E. Villamayor and D. Zelinsky**, Galois theory for rings with finitely many idempotents, *Nagoya Math. J.* **27**, 1966, 721–731

36. **O.E. Villamayor and D. Zelinsky**, Galois theory with infinitely many idempotents, *Nagoya Math. J.* **35**, 1969, 83–98

Index

F. Borceux, *Galois Theories of Fields and Rings*, Coimbra Mathematical Texts 2,
https://doi.org/10.1007/978-3-031-58460-2

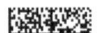